EAI/Springer Innovations in Communication and Computing

Series editor

Imrich Chlamtac, European Alliance for Innovation, Ghent, Belgium

Editor's Note

The impact of information technologies is creating a new world yet not fully understood. The extent and speed of economic, life style and social changes already perceived in everyday life is hard to estimate without understanding the technological driving forces behind it. This series presents contributed volumes featuring the latest research and development in the various information engineering technologies that play a key role in this process.

The range of topics, focusing primarily on communications and computing engineering include, but are not limited to, wireless networks; mobile communication; design and learning; gaming; interaction; e-health and pervasive healthcare; energy management; smart grids; internet of things; cognitive radio networks; computation; cloud computing; ubiquitous connectivity, and in mode general smart living, smart cities, Internet of Things and more. The series publishes a combination of expanded papers selected from hosted and sponsored European Alliance for Innovation (EAI) conferences that present cutting edge, global research as well as provide new perspectives on traditional related engineering fields. This content, complemented with open calls for contribution of book titles and individual chapters, together maintain Springer's and EAI's high standards of academic excellence. The audience for the books consists of researchers, industry professionals, advanced level students as well as practitioners in related fields of activity include information and communication specialists, security experts, economists, urban planners, doctors, and in general representatives in all those walks of life affected ad contributing to the information revolution.

About EAI

EAI is a grassroots member organization initiated through cooperation between businesses, public, private and government organizations to address the global challenges of Europe's future competitiveness and link the European Research community with its counterparts around the globe. EAI reaches out to hundreds of thousands of individual subscribers on all continents and collaborates with an institutional member base including Fortune 500 companies, government organizations, and educational institutions, provide a free research and innovation platform.

Through its open free membership model EAI promotes a new research and innovation culture based on collaboration, connectivity and recognition of excellence by community.

More information about this series at http://www.springer.com/series/15427

Raman Kumar • Sara Paiva
Editors

Applications in Ubiquitous Computing

RESEARCH MEETS INNOVATION

Editors
Raman Kumar
Department of Computer Science
and Engineering
I. K. Gujral Punjab Technical University
Kapurthala, Punjab, India

Sara Paiva
Inst Politécnico de Viana do Castelo
Viana do Castelo, Portugal

ISSN 2522-8595 ISSN 2522-8609 (electronic)
EAI/Springer Innovations in Communication and Computing
ISBN 978-3-030-35282-0 ISBN 978-3-030-35280-6 (eBook)
https://doi.org/10.1007/978-3-030-35280-6

This Springer imprint is published by the registered company Springer Nature Switzerland AG.
The registered company address is: Gewerbestrasse 11, 6330 Cham, Switzerland

Preface

In recent years, there has been an increasing focus on applications in ubiquitous computing including the theories, methodologies, and techniques underlying this evolving field as well as its potential use in various domains across the entire spectrum of sciences (natural science, health science, engineering, social science, and humanities) and in various types of businesses. Applications in the field of ubiquitous computing have yet to achieve their full potential but are projected to play a major role in the development of successful future advances in ubiquitous systems.

Applications in Ubiquitous Computing is an edited book that contains contributions from various experienced professionals. Topics range from the foundations of ubiquitous computing to examples of its use across multiple disciplines. The book enables scientists, scholars, engineers, professionals, policy-makers, and government and nongovernment organizations to share new developments in theory, analytical and numerical simulation and modelling, experimentation, demonstration, advanced deployment and case studies, results of laboratory or field operational tests, and ongoing developments with relevance to advances in ubiquitous computing.

The book is divided into 2 parts comprising of 12 chapters.

The first part includes chapters that primarily focus on the applications in ubiquitous computing, and the second part includes chapters that primarily demonstrate, through detailed studies and cases, how advances in ubiquitous computing can be used (and is already in use).

This book can benefit researchers, advanced students, as well as practitioners. The collection of papers in the book can inspire future researchers – in particular, researchers interested in interdisciplinary research. The rich interdisciplinary contents of the book can be of interest to faculty, research communities, and researchers and practitioners from diverse disciplines who aspire to create new and innovative research initiatives and applications. The book aims to inspire researchers and practitioners from different research backgrounds regarding new research directions and application domains within ubiquitous computing.

We wish to thank all the people who contributed to this edited book – the authors for their insightful contributions, the reviewers for their suggestions that ensured the quality of the individual parts, and, last but not least, the EAI/Springer team for their continuous support throughout the project. Without this joint effort, this book would not have been possible.

Kapurthala, Punjab, India   Raman Kumar
Viana do Castelo, Portugal   Sara Paiva

Contents

Part I
Ubiquitous Applications

The Pivotal Role of Internet of Things and Ubiquitous Computing in Healthcare

R. Udendhran and G. Yamini

1 Explosive Growth of Internet of Things and Ubiquitous Computing

IoT and ubiquitous computing in healthcare can therefore be considered an important life-saving technology in the healthcare domain, which is predominantly employed for gathering data from the bedside devices, viewing patient information, as well as diagnosing in real time the entire system of patient care [1].

Ubiquitous Computing (UbiComp) is characterized by the use of small, networked and portable computer products in the form of smart phones, personal digital assistants and embedded computers built into many devices, resulting in a world in which each person owns and uses many computers [2]. The consequences include enhanced computing by making computers available throughout everyone's daily life while those computers themselves and their interaction are 'invisible' to the users. The term 'invisible' in this context is used to mean the interaction between the computer and the user in a more natural manner such as speech and physical interaction, with the computer itself automatically capturing its external parameters while concurrently communicating with other computers [3]. UbiComp proposes many minute, wireless computers that can monitor their environments, and communicate and react to monitored parameters. However, the challenge which prevails in healthcare is that it causes loss of data and even fault in diagnosis and most of the confidential healthcare data are stored in cloud [4]. Even a minute in treating patients is life-saving; therefore doctors should save precious minutes,

R. Udendhran (✉) · G. Yamini
Department of Computer Science and Engineering, Bharathidasan University, Trichy, India

© Springer Nature Switzerland AG 2021
R. Kumar, S. Paiva (eds.), *Applications in Ubiquitous Computing*, EAI/Springer Innovations in Communication and Computing,
https://doi.org/10.1007/978-3-030-35280-6_1

3

which can be done by employing the monitoring of medical assets and less manual visiting each patient through remote diagnosis IoT and ubiquitous computing enabled remote diagnosis [5].

The advancements in healthcare are not implemented in developing countries because of the poor healthcare infrastructure [6].

The solution for this problem is to integrate health-sensing devices with portable devices such as smart phones and deploy them in the cloud. By following this technique, poor people can make use of healthcare by employing smart phones which are cheap these days [7].

However, healthcare should enhance the reliability by deploying real-time monitoring for patients and analysing patient data by providing smart healthcare monitoring devices since IoT has turned into an effective communication paradigm. Most of our daily life has become part of the internet because of its faster communication as well as capabilities. It is estimated that by 2020, more than 90% of the healthcare industry will integrate the IoT technology, which will in return enhance the efficiency of healthcare and provide quality care in the modern society [8–10]. Another compelling reason for adapting IoT technology in healthcare is that there is increase in the number of patients, which leads to a smaller number of doctors. Hence, most of the diagnostics are delayed because of this reason since it is time-consuming and even patients ignore diagnostics because of the expense and rely on the doctors. Many health problems are not detected because of non-availability of doctors as well as not accessing the healthcare systems. The only solution to this problem is to integrate healthcare with the IoT for real-time monitoring of every patient as well as analysing the data and providing real-time healthcare [11, 12].

Generally, the sensors employed in healthcare are used for real-time monitoring of patients and such devices are termed IoT-driven sensors [13, 14]. These types of sensors are deployed for patients in serious condition because of the non-invasive monitoring; for instance, the physiological status of patients will be monitored by the IoT-driven sensors that gather physiological information regarding the patient through gateways. This information is later analysed by the doctors and then stored in cloud, which enhances the quality of healthcare and lessens the cost burden to the patient [15]. The working principle of IoT in remote health monitoring systems is that it tracks the vital signs of the patient in real time and if the vital signs are abnormal, then it acts based on the problem in the patient and notifies the doctor for further analysis [16]. The IoT-driven sensor is attached to the patient, which transmits the data regarding the vital signs from the patient's location and employing a telecom network with a transmitter to a hospital which consists of a remote monitoring system that reads the incoming data about the patient's vital signs. In some cases, the sensor will be implanted into patient's body which transmits data electronically [17]. This confidential information will be encrypted and then decrypted for further analysis when the need arises [18].

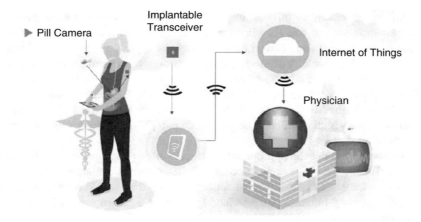

Fig. 1 Deployment of ubiquitous computing and IoT-driven sensors with oxygen saturation monitoring

2 Integrating Ubiquitous Computing and IoT-Driven Sensors with Rehabilitation Systems

A community-based smart rehabilitation system is employed for sharpening as well as rejuvenating the functional abilities that enhance the physiological care [19]. These systems should interact conveniently apportioning the medical resources based on the requirements of the patient employing the IoT-based rehabilitation system with an ontology-based automating technique, as shown in Fig. 1.

The pulse oximeter or an oxygen saturation device is used to monitor and analyse the patient's oxygen saturation following a non-invasive pattern [20] [21]. These devices were costly before the advent of IoT devices which led to many advancements in medical sensors as well as wireless networks because of less consumption of power [22]. The IoT-based pulse oximeter is deployed for medical applications that determine the blood oxygen levels as well as the heart rate with the help of sensors connected to patient's body [23].

3 Research Gap in Integrating Ubiquitous Computing and Intelligent Healthcare

So far, it has been said that the combination of healthcare with the information system falls under the category of healthcare information system. Moreover, in the present scenario, certain domains, such as Ubiquitous Computing, Big Data, Cloud Computing, M-Commerce, etc., also play a major role in this field. They form an integrated part of the healthcare organization systems [24]. Various studies had adapted the healthcare information systems. Several researchers have followed the information system and they had developed this information system in many healthcare institutions and showed their benefits to the wide range of stakeholders [25]. The basic functionality of this information system was explained in [26], along with the implementation and the adoption phase.

However, only a limited number of researches had focused on the selection of healthcare information system (HIS) that seems to be in the initial stage of implementation, which is found to be a research gap [27]. The determined research gap is mentioned in Fig. 2.

It is essential to select an appropriate healthcare information system, as shown in the figure. If little care is taken in choosing the healthcare information system, then as a result it may not fit the particular institution and they would have to perform certain unnecessary functions such as having resource waste in the institution with the decrease in efficiency [28].

There are various classifications under the healthcare information system, and it could be very easy to choose a classification system based on our requirement. However, the situation is the other way around because there is no unique categorization of healthcare information system [29]. But the categories have various features that could differ based on the categorization that could make the selection process a little complicated [30]. On the other hand, the classification could not be applied fully to the HIS since they could require certain features from the modern systems and most of them could be integrated and modular by these varying systems. At certain times, problems could occur in choosing the system, which is said to be known as the zoo problem. The problem that occurs from choosing an option from the wide variety

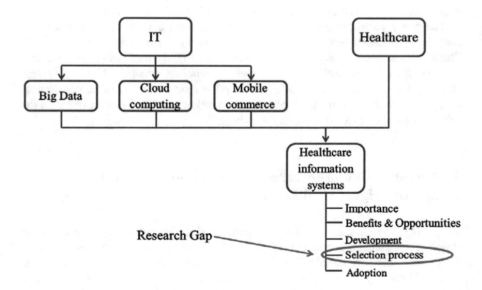

Fig. 2 Research gap in the information system

is said to be known as the zoo problem. This problem is very common among the users for whom they seem to be similar. Among 240 developers, there are around 650 systems, which seemed to be a complicated one in the selection process of Healthcare. Since all the options are found to be unique, the decision-makers tend to get confused, so they could choose an unsuitable information system [30].

4 Benefits of Cloud Computing in Healthcare

Several benefits could be attained by integrating cloud computing with the medical sector that drives the technology. Though there are certain difficulties, certain developing countries like Malaysia, Ghana and South Africa have witnessed improvement in healthcare with the cloud computing technology [31]. To enhance the outcomes of the patients, the Cloud technology embraces the healthcare that forms a great answer to all healthcare organizations. The interaction takes place among the patients, doctors and the Information and Communications Technology (ICT) environment through the internet to the cloud [32]. This process starts with the patients and their visit to the doctor who makes a verification of the patient's details in the eHealth (Cloud) system. The doctor analyses the patient's health with the details provided and offers certain medicines for their health with the possible medication. Through this system, the interactions could be very quick and the details of the patients could once again be updated in the cloud environment [33]. In the world of healthcare, cloud computing is said to be an emerging technology and seems to be very effective, when compared to the current health-based information storage systems [34].

The quick development of the device, wireless and cloud computing technologies alters a wise healthcare that supports the consistent remote watching on the physical conditions of patients, older individuals or babies, and the efficient process of the massive sensing knowledge sets. Such a smart healthcare will enhance the standard of life significantly [35]. However, as investigated by the Cloud Standards Customer Council, the healthcare institutions do not seem to be keen on building sensible healthcare systems supported by the IT, particularly in developing countries. The underutilization of IT prevents the wide information sharing and processing in the healthcare industry [36]. Cloud computing paradigms alter a virtual mechanism for IT resource management and usage. Sensor-cloud infrastructure could be a technology that integrates cloud techniques into the wireless sensor networks (WSNs). It provides users a virtual platform for utilizing the physical sensors in a clear and convenient approach. Users will manage, monitor, create and check the distributed physical sensors while not knowing their physical details, however simply using a few of functions. Cloud computing provides the IT resources for WSNs that support the storage and quick process of a large amount of detector information streams. The connection between the WSNs and clouds is enforced by two kinds of gateways: sensor gateways and cloud gateways, where sensor gateways collect and compress data, and cloud gateways decompress and process data.

Sensor-clouds are utilized in various applications, like the disaster prediction, atmosphere watching and healthcare analysis. Sensor-clouds collect knowledge from totally different applications and share and process the information based on the cloud computational and storage resources. An interface is provided to the users to manage and monitor the virtual sensors. In Sensor-clouds, the sensor modelling language (SML) is utilized to explain the data of the physical sensors which are processed as the metadata of the sensors. The standard language format allows the collaboration among sensors in several networks and platforms. The Extensible Markup Language (XML) encryption additionally provides a mapping mechanism for transforming commands and information between virtual and physical sensors.

Compared with the traditional sensor networks, Sensor-clouds have the following advantages: (1) the capability of dealing with numerous data types; (2) scalable resources; (3) user and network collaboration; (4) data visualization; (5) data access and resource usage; (6) low cost; (7) automated resource delivery and data management; (8) less process and response time.

5 IoT and Ubiquitous Computing Is Expected to Drive Growth in Intelligent Healthcare Information

Based on the effectiveness of IoT and Ubiquitous computing, an intelligent information system is said to be the set of software and hardware that involves the skilled people for the process of decision-making and co-ordination among the organizations. One of the recent concepts in decision-making involves the use of information system since it increases the facility of exchange of data among the users. The information system is used by the people and the organization to organize process, collect and distribute data. Through this process, quality information could be received that helps in taking the rational decisions in order to meet the needs of the customers.

Rational decision-making is said to be a crucial process unlike several other sectors, which offers certain provisions to the community. Moreover, fast increase in the medical data leads to the categorization of healthcare from the discovery of drugs, evolution and increase in the spread of diseases, medical histories, exposure of patients to the environment, etc. These fast-growing data are necessary to have an effective management to increase the quality of health service delivery. The increase in medical data had led to several challenges that had been faced by the hospitals in managing the information systems, which is sufficiently necessary to utilize the medical data for the rational decision-making process. The HIS had been mainly designed in governing the operation carried out in the healthcare organizations such as the clinic, patient registration, activities regarding the administration, financial aspects and communication. There are several categorizations under the HIS that are totally combined in managing the hospital administration and patient care. In this system, healthcare data storage with the information analysis is done systematically.

These data collected by the HIS are classified into three main categories: (1) clinical information that is collected during the patient's visit to the hospital, (2) administrative information and (3) external information. These include the data regarding the health facilities, household surveys and administrative data, national health accounts (NHA) and health researches and civil registrations. This could have access to the health information system by describing the situations and trends followed in the healthcare unit. Further, these situations could be thoroughly analysed and could be helpful in planning facility, co-ordination and in decision-making, as shown in Fig. 3.

Based on the type of the data and the purpose, the Healthcare Information System (HIS) could be classified into two categories: one is the clinical HIS and the other is the administrative HIS. On one side, it could be said that the administrative HIS will help in supporting the management and the general activities of the healthcare organization. It contains information regarding the finance, supplies as well as the human source, which also contains the financial management system with the patients' administrative system.

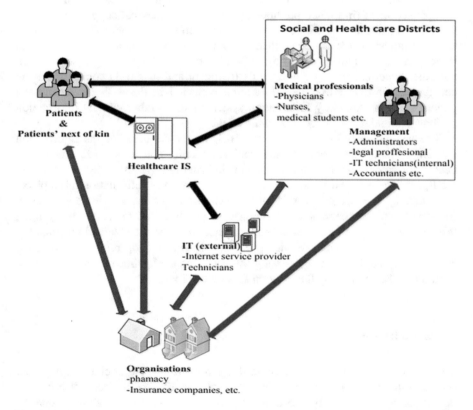

Fig. 3 Interaction among the healthcare actors and with the healthcare information system (HIS)

The employees and the patients are the important stakeholders or the participants of the healthcare system. The employee category has been classified into two main categories that include the administrative professionals as well as the medical professionals. The actors of the healthcare mainly consist of the organization or the individual, which gets an effect through the information systems. In this case, the individuals are said to be the people who have a connection with the information systems, for instance, administrators, clinicians, clinical students and the technologists. These kinds of people tend to work in various organizations such as the hospitals, pharmacy and a unit under organizations say, dispensary. Individuals Groups Organizational Human Controllers Acceptors Providers Supporters (IGOHcaps) is said to be a unit under the healthcare actors. The actors under the IGOHcaps could be classified into four main groups including Providers, Acceptors, Controllers and Suppliers. The combination of the healthcare actors is mainly appreciated, rather than separating these actors into various groups individually. Moreover, the healthcare industry supply chain also includes several entities such as the insurance providers, specialists, pharmacies, physicians and the medical suppliers. Further classification of these actors tends to be based on three main groups: suppliers (insurance providers, pharmacy and medical suppliers), patients and physicians. This has been composed by two different authors and finally it says that a healthcare actor can be defined as a human or organization that participates in accepting, providing, supporting or/and controlling the healthcare service. On the other hand, supporting the patient's treatment and diagnosis comes under the category of clinical HIS. The information collected in this HIS deals with the patient's health. Ancillary information systems and all other clinical information systems fall under the category of clinical HIS.

The healthcare information system (HIS) forms the base for the information exchange among the actors of the healthcare unit. The interaction among the healthcare actors and with the healthcare information system (HIS) is shown in the figure. There are two ways by which the exchange of information takes place: (1) either by the patients who visit the healthcare service providers physically (pharmacy, medical centres and insurance) or (2) they tend to search data online, take appointments and get the medical guidance based on the obtained information. The ICT for the healthcare service providers should function properly during the process information exchange among the actors of the healthcare and this aspect should be taken care of by the external IT service providers.

6 Conclusion

The quality of healthcare is necessarily important for the social stability and economic impact on the citizens. The efficiency, quality and the cost-effectiveness of the service seems to be the most important sector for the healthcare quality. Liberating the society from a disease-free environment is really complimented and therefore the cost of an effective technology in healthcare could be complicated.

People who cannot afford to access the healthcare techniques could face several contagious diseases. In order to avoid this situation, relevant choice of a cost-efficient Information Technology (IT) platform is necessary. The use of Internet of things and Ubiquitous computing in the field of healthcare seems to increase the simplification and thereby enhance the provisions and the quality of healthcare.

References

1. Xu, B., Xu, L. D., Cai, H., Xie, C., Hu, J., & Bu, F. (2014, May). Ubiquitous data accessing method in IoT-based information system for emergency medical services. *IEEE Transactions Industrial Informatics, 10*(2), 1578–1586.
2. Olaronke, I., & Oluwaseun, O. (2016, Dec). Big data in healthcare: Prospects, challenges and resolutions. *Proceedings of the Future Technologies Conference (FTC)*, 1152–1157.
3. Zhou, J., Cao, Z., Dong, X., & Vasilakos, A. V. (Jan. 2017). Security and privacy for cloud-based IoT: Challenges. *IEEE Communications Magazine, 55*(1), 26–33.
4. Symptoms of Parkinson's. (2017). [online] Available: https://shakeitup.org.au/understanding-parkinsons/symptoms-of-parkinsons/
5. Ženko, J., Kos, M., & Kramberger, I. (2016, May). Pulse rate variability and blood oxidation content identification using miniature wearable wrist device. *Proceedings of the International Conference on Systems, Signals and Image Processing (IWSSIP)*, 1–4.
6. HSM-Tri. (2017). [online] Available: https://buy.garmin.com/en-AU/AU/p/136403
7. H7 Heart Rate Sensor. (2017). [online] Available: https://www.polar.com/au-en/products/accessories/H7_heart_rate_sensor
8. FitBit PurePulse. (2017). [online] Available: https://www.fitbit.com/au/purepulse
9. Lee, H., Ko, H., Jeong, C., & Lee, J. (2017, Feb). Wearable photoplethysmographic sensor based on different LED light intensities. *IEEE Sensors Journal, 17*(3), 587–588.
10. Shu, Y., Li, C., Wang, Z., Mi, W., Li, Y., & Ren, T.-L. (2015). A pressure sensing system for heart rate monitoring with polymer-based pressure sensors and an anti-interference post processing circuit. *Sensors, 15*(2), 3224–3235.
11. Wang, D., Zhang, D., & Lu, G. (2015, Jul). A novel multichannel wrist pulse system with different sensor arrays. *IEEE Transactions on Instrumentation and Measurement, 64*(7), 2020–2034.
12. Wang, D., Zhang, D., & Lu, G. (2016, Mar). An optimal pulse system design by multichannel sensors fusion. *IEEE Journal of Biomedical and Health Informatics, 20*(2), 450–459.
13. Zuo, W., Wang, P., & Zhang, D. (2016, Jan). Comparison of three different types of wrist pulse signals by their physical meanings and diagnosis performance. *IEEE Journal of Biomedical and Health Informatics, 20*(1), 119–127.
14. An, Y.-J., Kim, B.-H., Yun, G.-H., Kim, S.-W., Hong, S.-B., & Yook, J.-G. (2016, Apr). Flexible non-constrained RF wrist pulse detection sensor based on array resonators. *IEEE Transactions on Biomedical Circuits and Systems, 10*(2), 300–308.
15. Milici, S., Lorenzo, J., Lázaro, A., Villarino, R., & Girbau, D. (2017, Mar). Wireless breathing sensor based on wearable modulated frequency selective surface. *IEEE Sensors Journal, 17*(5), 1285–1292.
16. Varon, C., Caicedo, A., Testelmans, D., Buyse, B., & van Huffel, S. (2015, Sept). A novel algorithm for the automatic detection of sleep apnea from single-lead ECG. *IEEE Transactions on Biomedical Engineering, 62*(9), 2269–2278.
17. Oletic, D., & Bilas, V. (2016, Dec). Energy-efficient respiratory sounds sensing for personal mobile asthma monitoring. *IEEE Sensors Journal, 16*(23), 8295–8303.
18. Yang, X., et al. (2015, Feb). Textile fiber optic microbend sensor used for heartbeat and respiration monitoring. *IEEE Sensors Journal, 15*(2), 757–761.

19. Min, S. D., Yun, Y., & Shin, H. (2014, Sept). Simplified structural textile respiration sensor based on capacitive pressure sensing method. *IEEE Sensors Journal, 14*(9), 3245–3251.
20. Mahbub, I., et al. (2017, Mar). A low-power wireless piezoelectric sensor-based respiration monitoring system realized in CMOS process. *IEEE Sensors Journal, 17*(6), 1858–1864.
21. Atalay, O., Kennon, W. R., & Demirok, E. (2015, Jan). Weft-knitted strain sensor for monitoring respiratory rate and its electro-mechanical modeling. *IEEE Sensors Journal, 15*(1), 110–122.
22. Aqueveque, P., Gutiérrez, C., Rodríguez, F. S., Pino, E. J., Morales, A., & Wiechmann, E. P. (2017, May/Jun). Monitoring physiological variables of mining workers at high altitude. *IEEE Transactions on Industry Applications, 53*(3), 2628–2634.
23. Narczyk, P., Siwiec, K., & Pleskacz, W. A. (2016, Apr). *Precision human body temperature measurement based on thermistor sensor*. Proceedings of the 2016 IEEE 19th International Symposium on Design and Diagnostics of Electronic Circuits & Systems (DDECS) (pp. 1–5).
24. Nakamura, T., et al. (2016, Feb). *Development of flexible and wide-range polymer-based temperature sensor for human bodies*. Proceedings of IEEE-EMBS International Conference on Biomedical and Health Informatics (BHI) (pp. 485–488).
25. Gope, P., & Hwang, T. (2016). BSN-care: A secure IoT-based modern healthcare system using body sensor network. *IEEE Sensors Journal, 16*(5), 1368–1376.
26. Yeh, K. H. (2016). A secure IoT-based healthcare system with body sensor networks. *IEEE Access, 4*, 10288–10299.
27. Kern, S. E., & Jaron, D. (2003, Jan-Feb). Healthcare technology, economics and policy: An evolving balance. *IEEE Engineering in Medicine and Biology Magazine, 22*, 16–19.
28. Wells PNT. (2003, Jan–Feb). Can technology truly reduce healthcare costs. *IEEE Engineering in Medicine and Biology Magazine, 22*, 20–25.
29. Suresh, A. (2017). Heart disease prediction system using ANN, RBF and CBR. *International Journal of Pure and Applied Mathematics, (IJPAM), 117*(21), 199–216. ISSN: 1311-8080, E-ISSN: 1314-3395.
30. Yuan, B., & Herbert, J. (2012). Fuzzy CARA – A fuzzy-based context reasoning system for pervasive healthcare. *Procedia Computer Science (ANT), 10*, 357–365.
31. Sun, H. M. (2014). Online smoothness with dropping partial data based on advanced video coding stream. *Multimedia Tools and Applications, 69*, 1021. https://doi.org/10.1007/s11042-012-1141-x.
32. Suresh, A., Kumar, R., & Varatharajan, R. (2018). Health care data analysis using evolutionary algorithm. *The Journal of Supercomputing*. https://doi.org/10.1007/s11227-018-2302-0.
33. Vijayalakshmi, K., Uma, S., Bhuvanya, R., & Suresh, A. (2018, Feb). A demand for wearable devices in health care. *International Journal of Engineering and Technology, 7*(1.7), 01–04. https://doi.org/10.14419/ijet.v7i1.7.9377. ISSN: 2227-524X.
34. Udendhran, R. (2017, Mar 22–23). *A hybrid approach to enhance data security in cloud storage*. ICC'17 Proceedings of the Second International Conference on Internet of things and Cloud Computing at Cambridge University, UK. ISBN: 978-1-4503-4774-7. https://doi.org/10.1145/3018896.3025138.
35. Suresh, A., Udendhran, R., Balamurgan, M., et al. (2019). A novel internet of things framework integrated with real time monitoring for intelligent healthcare environment. *Springer-Journal of Medical System, 43*, −165. https://doi.org/10.1007/s10916-019-1302-9.
36. Suresh, A., Udendhran, R., & Balamurgan, M. (2019). Hybridized neural network and decision tree based classifier for prognostic decision making in breast cancers. *Springer - Journal of Soft Computing*. https://doi.org/10.1007/s00500-019-04066-4.

Easy to Read (E2R) and Access for All (A4A): A Step to Determine the Understandability and Accessibility of Websites

Ranjit Singh and Abid Ismail

1 Introduction

World Wide Web (WWW) is one of the important applications of the Internet that removes the physical barrier to access services, products, and information that are not easily obtained by PwDs because of circumstances related to their disability for independent living and enhances their decision-making ability [60, 84]. It is no doubt that web is an important "able-bodied" neighbor or coworker, if you are a PwD [55]. Following the development of commercial applications, many researchers [9, 15, 28, 68, 71] have realized the potential of the Internet and WWW in tourism business and recommended for incorporating them into the travel and tourism industry. The incorporation of the Internet has revolutionized the travel and tourism business and is also significant for the growth and success of the industry [11, 13, 29, 62]. It is still continued.

Tourism is an information consumption industry, and WWW is able to serve that information over the Internet to the user for decision-making regarding tourism [9, 10, 59, 62]. With the development of Web 2.0, online users are empowered through technology [52, 53]. It is a tool of mass collaboration because it allows users to actively participate and collaborate with other users to produce, consume and diffuse the knowledge and information on tourism being distributed over the Internet [61, 65, 66]. The online travel agencies (OTAs) are one of the applications of the Internet in the tourism industry, and their websites hold the majority share of

R. Singh
Department of Tourism Studies, School of Management, Pondicherry University, Puducherry, India

A. Ismail (✉)
Department of Computer Science, School of Engineering and Technology, Pondicherry University, Puducherry, India

© Springer Nature Switzerland AG 2021
R. Kumar, S. Paiva (eds.), *Applications in Ubiquitous Computing*, EAI/Springer Innovations in Communication and Computing,
https://doi.org/10.1007/978-3-030-35280-6_2

the travel agency business and continue to witness huge growth by 2025 [86]. They are making the travel simple and personalized for everyone. The major advantage of OTAs is booking an entire trip in one sitting, finding the best bargaining price, discovering travel inspiration, getting a glimpse of your travel, and sealing the deal anytime [58]. Online travel market places become more competitive due to the entry of new OTAs in the market and are increasingly adopting e-business model to accomplish their organizational goals [42]. Therefore, maintaining an effective and customized website has thus become imperative for OTAs to strengthen its customer relationships, win a larger market segment and serve small niche market segment like accessible tourism.

Website of OTAs is one of the dominant sources of information for the accessible tourism consumers with different access needs which empowers the disabled people to search for, find, plan, compare, bargain, book travel, and tourism experiences online and give feedback after using the products. Accessible tourism consumers are a growing niche market segment of tourism with full of opportunities and challenges. As per findings, the average yearly expenditure of consumers on tourism is EUR 80 billion in Europe, USD 13.6 billion in the USA, and AUD 1.3 billion in Australia [76]. It shows the potential of accessible tourism to contribute toward the economy. However, the market is underperforming due to different types of environmental barriers (planning and booking, infrastructure and transportation, building, communication and activities involving destination) and social barriers (lack of awareness about accessibility, lack of training, and tourism-related business and attitudinal barriers) [76]. From the aforementioned barriers, lack of access to information through the website is a major barrier for accessible tourism because information is the lifeblood of tourism [62]. Most of the tourism websites have failed to address the issue of information accessibility through their website toward PwDs [22, 79].

Understanding the needs of the online users with or without disability translates to the success of tourism websites, and it is of utmost importance to the tourism and hospitality organization [41]. Accessibility and readability issues in tourism websites are imperative for both travel and tourism industry and accessible tourism consumers. Accessible OTA websites can make a difference in the highly competitive tourism market as such websites will entice more customers, especially PwDs, providing better opportunities to create direct relationships with the consumer and loyalty in the long run for the OTAs. However, the accessibility and readability of OTA websites are still questionable and yet to unearth due to the importance of OTAs in accessible tourism. Hence, it is imperative to examine the accessibility and readability of OTA websites toward the commitment of accessible tourism.

Thus, the aim of this study is to provide an in-depth understanding of web accessibility and web readability toward accessible tourism from the World Wide Web Consortium (W3C) guideline perspective. For this, web accessibility of 35 OTA websites belonging to three international corporations has been checked by applying online open access tools, namely, *AChecker, WAVE, and Tenon*, and their understandability performance through six different readability indices. In addition, ranking has been performed in terms of accessibility and readability violation scores along with site visiting ranking by *Alexa ranking tool*. Therefore, this research finding would be of great interest to many organizations to embody an open, welcoming, and inclusive environment for PwDs to carry out optimal tourism experiences and boost their universality of the web for all.

2　Objectives

The following are the main objectives of this study:

1. To determine the web accessibility score of 35 OTA websites in terms of WCAG 2.0 guidelines.
2. To produce the readability score of OTA websites by applying six different indices.
3. To describe the relevance of web accessibility and readability of OTA websites on accessible tourism.
4. To rank the websites based on the accessibility and readability score and compare the rank with site visiting ranking.
5. To check performance among the three international corporations in terms of testing techniques used.

3　Literature Review

3.1　Online Travel Agency

The Internet has been widely accepted by the tourism researchers that it can serve a valuable tool for the promotion and distribution of travel- and tourism-related experience of customers [11, 16, 17, 70, 87].

It is a valuable online platform for both consumers and suppliers for the dissemination of information, communication, and online purchasing of product and service related to the destination without any geographical and time constraint [42, 48, 72]. Websites have developed into one of the leading points of supply of information for tourism and hospitality, since they produce details about transportations, holiday packages, and hotels of a destination. Due to the growing use of websites for purchasing travel-related products, the online travel industry is forecasted to reach $1,091 billion by 2022, and the major booking source for the user includes websites of direct travel suppliers and websites of OTAs [7].

Tourism academic researcher has added accessibility is an important criteria of OTAs for service quality (E-QUAL) to improve customer satisfaction and loyalty in the travel and tourism industry [39]. More importantly, accessibility is a critical criteria to measure the quality of websites for OTAs [38, 49]. High-quality websites can entice more consumers than low-quality websites [56, 85, 94].

Considering the importance of accessibility of OTA websites in travel planning and booking, it should be expected from the OTAs that equal chances and opportunity must be met especially by PwDs due to their access needs without any digital divide. Nevertheless, to be easy to access those features, it would be reasonable that PwDs should be able to access the e-distribution, an issue that has not been addressed by travel and tourism industry as a whole [46].

The Internet has made the users independent in planning their trips, since it provides updated and detailed information related to a destination for their decision-making by sitting at one place [10]. In the recent two decades, the quantity of studies on the importance of information for PwDs has increased [12, 18, 24]; however, no further understanding has been achieved into the significance of OTA websites from functionality perspective, that is, evaluating the accessibility and readability of the website. Therefore, it is necessary to research on the readability and accessibility of OTA websites, to remove the barriers by providing accessible information of the tourism products readily available.

3.2 Accessible Tourism

More than a billion of people are estimated to experience some form of disability, and the prevalence of disability is high in developing countries [92]. The number is going to increase because population are aging and there is a global increase in chronic health issues associated with disability, such as cardiovascular diseases, diabetes, and mental illness. The International Classification of Functioning, Disability and Health (ICF) and the Convention on the Rights of Persons with Disabilities (CRPD) highlighted the major barriers of disability that are widespread in the environment which include inadequate policy and standard, negative attitude, lack of provision of service, problems with the service delivery, inadequate funding, lack of accessibility, lack of evidence and data, and lack of consultation and involvement [73, 92].

Due to the growing level of social integration and economic condition, PwDs are participating in tourism activities frequently [75]. Consequently, tourism researchers have reported that the market of accessible tourism has a big opportunity for the travel and tourism organizations having extensive growth and future possibilities [4, 8, 14, 19]. A substantial amount of research has been done to capture the accessible tourism market and its barrier from both demand and supply sides of tourism [6, 57, 67]. Although different steps are taken to make tourism accessible for everyone especially PwDs, access to information through websites is a prominent issue in the travel and tourism industry for disabled people [22, 89]. Majority of the tourism websites are not accessible and are not following the WCAG guideline [25, 46, 79]. Therefore, creating accessible websites should be one of the basic elements in the development of accessible tourism.

3.3   Web Accessibility

Web accessibility is defined as the websites, tools, and technologies that are designed and developed so that PwDs can *perceive, operate, navigate,* and *interact* with the web [83]. Hence, the effect of disability completely changed on the web as it eliminates barriers to communication that many people encounter in the physical world. However, when applications and websites are poorly designed, they can create barriers for the inclusion of people from using the website. Accessibility is important for organizations that want to create an optimal website and include people for using services and products. But it is rarely found in the tourism and hospitality industry [22, 63, 91].

Web accessibility for PwDs is a growing field of research in human-computer interaction [2, 20, 26, 35, 64]. In education and government field, a plethora of research on web accessibility has been conducted by different researchers due to its importance for disabled people [3, 30, 32–34, 36, 37]. They used different online automatic tools like AChecker, WAVE, TAW, Cynthia Says, readFX, etc. for the evaluation process of websites. However, in tourism scholarship marginal research has been conducted on web accessibility mostly confined to websites of destination marketing organizations (DMOs), websites of national tourist organizations (NTOs), and hotel websites based on WCAG 1.0, undermining the relevance of OTA websites in accessible tourism [22, 44, 50, 63, 89, 91]. Not only web accessibility but also web readability determines the success of the tourism websites for promoting accessible tourism, which is always disregarded by the tourism researchers [63].

Web accessibility is not a new concept in tourism and hospitality industry, but the research is limited mostly on the online tool Bobby with guideline WCAG 1.0 [27, 63, 91, 93]. For the web accessibility of US airline online reservation websites, when assessed with Bobby online, it was found that out of 73 websites, only 3 passed the initial test for accessibility, more than 75% sites contains 3 or more errors, and the most prominent error was alternative text to all images [27]. Not only US airline reservation websites but also Visitor Information Centers (VIC) websites of Queensland, Australia, fails to provide a text equivalent for each image in the main web page when assessed with priority level 1 of WCAG 1.0 through Bobby online [63]. Despite various rules and regulations, most of the developed countries' hotel websites such as in the UK, the USA, and Australia were inaccessible for PwDs and failed to one or more checkpoints of WCAG [46, 88, 90, 91]. In addition, in websites of tourism promotion organization, national airlines, lodging and hotels, tour operators, and travel agencies of developing countries such as in Uganda, South Africa, Kenya and Zimbabwe when tested by using LIFT and Bobby online with WCAG 1.0 guidelines, it was reported that 92% of websites were missing alternatives to visual and audio content and 67% of websites failed to address the issue of dynamic content in the website [44].

In 2004, researchers [89] tested the accessibility of 100 German and UK tourism-related websites and found that the home page of 10 UK and 10 German websites was barrier-free regarding priority 1 checkpoints, only 3 German websites passed the priority 2 checkpoints, and 2 UK and 1 German website passed the priority checkpoints of WCAG 1.0 [89]. Xiong et al. [93] expanded the accessibility test of websites by adding both WCAG 1.0 and Section 508 guidelines for measuring accessibility. The study found that websites had poor level of accessibility, and the majority of the websites failed in providing alternative text for the non-text element. Oertel et al. [50] assessed the accessibility of 16 official national tourist organization websites of the European Union. The research found that none of these websites met first priority of WCAG 1.0, and the websites even lack of basic and easy checkpoints, for instance, alternative text, clear navigation mechanism, etc. The official tourism website of Denmark was more accessible, and the website deliberately refrained from time-dependent elements.

Recently due to the development of online tools based on WCAG 2.0 and WCAG 2.1 guidelines, some research has measured the web accessibility of NTOs on the basis of WCAG 2.0 guidelines [21, 22, 79]. In the websites of official NTOs of countries registered in the United Nations World Tourism Organization (UNWTO), when tested by using TAW in terms of conformance level AA and AAA of WCAG 2.0, it was found that South Korea, Hong Kong, and Japan are following the good practice of accessibility as compared to other countries. They can serve as an example for other countries regarding accessibility [79]. Vila et al. [21] tested the accessibility of 190 websites of official NTOs around the world. The study found that the number of problems in success criteria was 2051, the number of warning in success criteria was 8096, and the number of not reviewed in success criteria was 188 for conformance level AA, and the number of problems in success criteria was 2038, the number of warning in success criteria was 6927, and the number of not reviewed in success criteria was 191 for conformance level AAA of WCAG 2.0 [21]. Besides this, another study is conducted to examine the web accessibility of Northern European countries' tourism websites by the same author, by applying the same TAW tool and conformance level AA of WCAG 2.0. The study reported 2319 total problems, 9644 total warnings, and 379 total not reviewed. The websites of Norway had the maximum number of incidents, and websites of Belgium had the least number of incidents [22].

3.4 Web Readability

Readability is how easily a reader can read and understand the words, sentences, written text, and style of writing [23]. It is based on the principles of legibility, familiarity, complexity, and typography of a sentence. Readability is measured through readability score by applying different readability indices available. Readability score tells the level of education required to understand a piece of text. The most commonly used readability formulas to measure readability include Flesch-Kincaid Reading Ease (FKRE), Flesch-Kincaid Grade Level (FKGL), Gunning Fog Score, Coleman-Liau Index (CLI), Automated Readability Index (ARI), and SMOG Index. These formulas test the readability based on the words, sentences, syllables, average sentence length, percentage of hard words, and characters in a sentence. The readability analysis of 75 Australian and New Zealand tourism websites was tested by using the above tools, and it was found out that it is an issue for the older age people to understand the text of the websites, and the grade level is high (difficult to read) [43]. Researchers [32, 34, 51], etc. used these aforementioned readability indices to find the grade level and understandability of texts in websites. Similarly, we have also used these readability indices in our study to find the grading cum understandability scores of 35 OTA websites.

Based on the review of literature, there is a need to focus on accessibility and readability of tourism websites regarding different corporations and to find their bonding strengths between the parameters of accessibility, readability, and site visiting status of OTA websites belonging to international corporations.

4   Methodology

4.1   Data Collection Method

Based on operation around the globe findings [69], three major US travel international corporations, namely, *Booking Holdings Inc.*,[1] *Expedia Group Inc.*,[2] and *TripAdvisor Inc.*,[3] are used in this study. A total of 35 OTA websites collected from each corporation having individual brands on their global website consists of 14, 16, and 4 websites of *Expedia Group, TripAdvisor,* and *Booking Holdings Incs.,* respectively.

4.2   Web Content Accessibility Guideline (WCAG)

WCAG is a technical document on standards of web accessibility developed by the *Accessibility Guidelines Working Group (AGWG)*.[4] AGWG is a part of the World Wide Web Consortium (W3C) Web Accessibility Initiative (WAI). W3C WAI develops support materials and standards to help organization and individual for understanding and implementing accessibility. W3C WAI resources are used to make applications, websites, and other digital creations more accessible and usable to everyone. W3C WAI combines people from disability organization, government, industry, and research laboratory from around the globe to develop resources and guidelines to make the web accessible to *people with neurological, cognitive,*

---

[1]https://www.bookingholdings.com/

[2]https://www.expediagroup.com/

[3]https://www.tripadvisor.com

[4]https://www.w3.org/WAI/GL/

auditory, speech, physical, and visual disabilities [83]. WCAG is developed in cooperation with organizations and individuals around the globe with an objective of creating and providing a standard of web content accessibility that satisfies the needs of governments, individuals, and organizations internationally. WCAG explains how to make web content more accessible to PwDs. Web content usually refers to the information in a web application or web page containing markup or code that defines presentation, structure, etc. and natural information such as images, text, and sounds. WCAG is primarily designed for web accessibility evaluation tools for developers, web authoring tool developers, web content developers, and others who need or want a standard for web accessibility including mobile accessibility. There are different versions of WCAG available, namely, *WCAG 1.0, WCAG 2.0, and WCAG 2.1.*

1. WCAG 1.0 was published in May 1999. It included 14 guidelines and 65 checkpoints. Each checkpoint has a priority (P) levels, namely, P1, P2, and P3, authorized by the working group on the basis of checkpoint's impact on accessibility [40]. In addition, there are three levels of conformance in WCAG 1.0 document including *Conformance Level "A"*, all P1 checkpoints are satisfied; *Conformance Level "AA"*, all P1 and P2 checkpoints are satisfied; and *Conformance Level "AAA"*, all P1, P2, and P3 checkpoints are satisfied [80].
2. WCAG 2.0 was published by W3C in December 2008. It succeeds WCAG 1.0. It was launched due to some shortfalls in WCAG 1.0 including some checkpoints that has been obsoleted because of development in technology for PwDs, moving from specific technology to technology independently, and reorganizing and improving accessibility guidelines to enhance accessibility [40]. It encompasses a wide range of suggestions and recommendations to make the web content and application more accessible. It will also address the issue of accessibility to a wide range of disabled people including learning disabilities, cognitive limitation, deafness and hear loss, limited movement, blindness and low vision, photosensitivity and combination of these. WCAG 2.0 not only improves the accessibility of the web content, but it also makes the web content more usable to users.

   WCAG 2.0 consists of *four* principles, namely, *perceivable, operable, understandable, and robust, denoted as POUR*. Under these principles, there are 12 guidelines that provide basic objectives to web content developers to make the web content more accessible to people with different disabilities. Under each guideline, success criteria are provided regarding three conformance levels: A (lowest), AA (medium), and AAA (highest) [81].
3. WCAG 2.1 is an extension of WCAG 2.0, published in December 2008 as a recommendation of W3C. Web content that conforms the guidelines of WCAG 2.1 also conforms the guidelines of WCAG 2.0 [82]. The publication of WCAG 2.1 is not to supersede WCAG 2.0, while W3C recommends WCAG 2.0. The W3C advises using standards of WCAG 2.1 to maximize the accessibility efforts and also encourages to use the current version of WCAG. WCAG 2.1 was initiated with an objective to enhance accessibility guidance for three major

groups, namely, *users with disabilities on mobile devices, users with cognitive or learning disabilities,* and *users with low vision.* It expands WCAG 2.0 by including new success criteria, definitions to support the success criteria, guidelines to organize the additional success criteria, and a couple of additions to the conformance section. This additional success criterion helps to make it clear that websites which conform to the success criteria of WCAG 2.1 also conform to the success criteria of WCAG 2.0 [82].

4.3   Tools and Techniques Used

There are different tools and techniques used for checking the accessibility status of websites based on the World Wide Web Consortium website.[5] These tools and techniques are based on different versions such as WCAG 1.0, 2.0, and 2.1, Section 508, etc. of accessibility guidelines and standards. Among them, some tools are open access and some are paid. But, we used open access tools based on WCAG 1.0 and 2.0, namely, *AChecker, WAVE, and Tenon*, for the evaluation process of 35 websites belonging to aforementioned three international corporations. The working snapshots of AChecker, WAVE, and Tenon tools are shown in Figs. 1, 2, and 3, respectively. In addition, the study used the online *WebFX tool*[6] to test the readability cum grade level means easy to read (E2R)

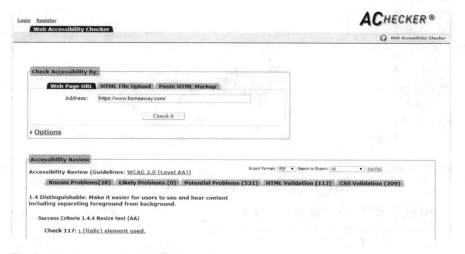

Fig. 1 A working snapshot of AChecker tool

---

[5]https://www.w3.org/WAI/ER/tools/

[6]https://www.webfx.com/tools/read-able/, a full-service digital market agency.

Fig. 2 A working snapshot
of WAVE tool

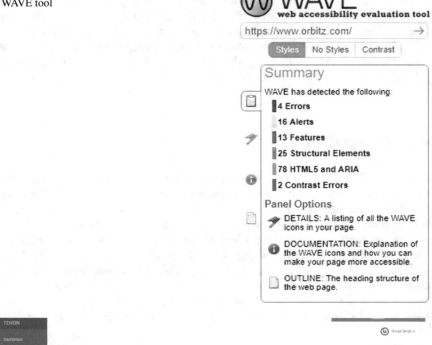

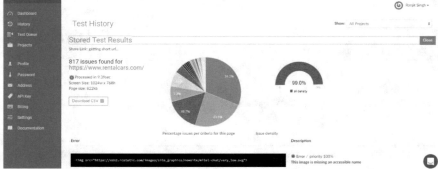

Fig. 3 A working snapshot of Tenon tool

scores of 35 OTA websites using *test by URL* technique. It includes six readability
index results, namely, *Flesch-Kincaid Reading Ease (FKRE), Flesch-Kincaid Grade
Level (FKGL), Gunning Fog (GFOG), Simple Measure of Gobbledygook (SMOG),
Coleman-Liau Index (CLI), and Automated Readability Index (ARI)*. All the six
readability indices involve different equations in terms of counting words, sentences,
syllables, complex words and sentences, etc. to measure the websites based on US
grading system of understandability. The working snapshot of WebFX tool is shown
in Fig. 4.

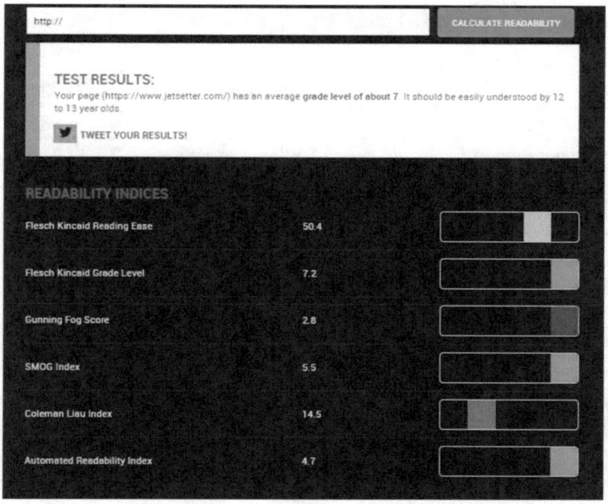

Fig. 4 A working snapshot of readability tool

Furthermore, to find the ranking of these 35 websites globally and nationally, we used online tool called *Alexa*.[7] In this study, we extract the global ranking of these aforementioned 35 websites which are later used to find the relevance with respect to accessibility and readability variables too. Also, SPSS procedure is used to find the correlations among the selected variables, namely, AChecker, WAVE, GFOG, and Alexa, called statistical analysis.

5 Result Analysis and Discussion

Based on the analysis of 35 websites, the following results shown in Table 1 with respect to access for all called accessibility (AChecker, WAVE, and Tenon), easy to read called readability (GFOG), and most visiting websites called ranking of Alexa (Alexa Global) are found.

5.1 Access for All Result

AChecker tool is used to test the selected 35 websites regarding conformance levels (A, AA, AAA), enabling HTML and CSS validators [1]. It identifies three types of problems.

---

[7]https://smallseotools.com/Alexa-rank-checker/

Table 1 Accessibility, readability, and ranking performance of 35 OTA websites

	Performance scores based on different methods					
S.No.	OTA website	AChecker	WAVE	GFOG	Alexa Global	Tenon
1	https://www.expedia.com/	1448	399	5.8	585	NA
2	https://www.hotels.com/	1600	196	4.2	715	480
3	https://www.vrbo.com/	2260	467	6.6	2310	921
4	https://www.trivago.com/	558	120	6.1	7414	22
5	https://www.homeaway.com/	2622	467	8.8	6796	1007
6	https://www.orbitz.com/	834	137	5.4	3862	NA
7	https://www.travelocity.com/	789	149	3.7	4170	NA
8	https://www.hotwire.com/	437	263	0.8	5291	2155
9	https://www.wotif.com/	812	150	4.5	31,447	NA
10	https://www.ebookers.com/	812	145	3.5	79,307	NA
11	https://www.cheaptickets.com/	279	137	5.3	17,618	NA
12	https://www.carrentals.com/	1486	228	6.5	41,146	588
13	https://www.cruiseshipcenters.com/	2340	582	2.6	256,414	1826
14	https://www.classicvacations.com/	5519	489	12.1	315,662	1437
15	https://www.tripadvisor.com/	2412	672	5.9	264	3585
16	https://www.airfarewatchdog.com/	2681	166	3.4	19,909	1869
17	https://www.bookingbuddy.com	248	106	9.8	20,727	255
18	https://www.familyvacationcritic.com/	2349	399	5.7	91,410	818
19	https://www.cruisecritic.com/	2543	186	7.4	11,932	623
20	https://www.flipkey.com/	2104	123	5.9	62,610	526
21	https://www.holidaylettings.co.uk/	2687	106	4.2	76,220	799
22	https://www.holidaywatchdog.com/	317	92	6.8	1,023,328	218
23	https://www.housetrip.com/	1750	223	5.6	416,668	594
24	https://www.jetsetter.com/	1735	368	2.8	45,876	916
25	https://www.oyster.com/	1343	113	4.2	21,598	295
26	https://www.seatguru.com/	2392	645	3.6	8102	5212
27	https://hotels.tingo.travel/	80	56	7.1	180,645	100
28	https://www.vacationrentals.com/	1685	121	4.8	152,508	424
29	https://www.viator.com/	2728	127	6.3	4544	1269
30	https://www.onetime.com/	167	113	9.7	400,337	390
31	https://www.booking.com/	3213	467	6.2	89	1451
32	https://www.kayak.com/	4559	1140	4.1	895	NA
33	https://www.priceline.com/	583	179	5.3	2200	NA
34	https://www.agoda.com/	1352	274	6.8	696	956
35	https://www.rentalcars.com/	6517	318	5.2	3320	817
Mean		1864.03	283.51	5.62	94,760.43	1094.56
STDev		1463.75	226.40	2.21	196,981.49	1120.58

1. *Known problems:* These are the accessibility barriers. OTA must modify these errors to make their web page accessible.

2. *Likely problems:* These are the probable errors and require human decision. OTA can modify these errors to increase the accessibility of their website.
3. *Potential problems:* AChecker cannot identify these problems and require human to take decision. OTA may modify the web page to address these problems or simply conform that the problem identified is not present.

Therefore, Table 2 presents the overall AChecker evaluation result of 35 websites based on 3 international corporations. It was found that their aggregate mean and standard deviation is more better than the individual level of conformance. Under level AA, potential problems are very high which need to be minimized. Based on the findings, the overall score should be minimized so that accessibility for all may be achieved.

After the analysis of AChecker, another accessibility tool is used called *WAVE*[8] to identify six types of problems, namely, errors, alerts, features, structural elements, HTML5 and ARIA, and contrast errors. It indicates that errors are accessibility errors that need to be addressed, while features are probable accessibility features that need to be addressed to improve the accessibility. The objective should not be to get rid of all the problems, except for the errors. Alerts require close examination, and other problems are displayed to facilitate human analysis of accessibility. The overall violation scores of accessibility among the selected 35 websites based on WAVE tool are shown in Fig. 5.

Using Tenon tool[9] to identify the worst performing web pages among the selected 35 websites, it is found that out of 35 websites, 17 are worst performing websites regarding total issues, errors, and warning densities which is shown in Table 3.

The most common identified issues regarding features among the websites are shown in Table 4. It is found that language, typography, and content, CSS, images and other non-text content, and navigation features are highly violated, having violation scores of 10,953, 6155, 3186, and 2874, respectively. It is highly suggested to web developers and designers to focus on these issues by content category to minimize them so that accessibility among the sites is attained in a more better way.

Table 2 AChecker tool evaluation result of 35 websites based on 3 international corporations

AChecker tool report							
35 OTA websites	Known problems	Likely problems	Potential problems	HTML validation	CCS validation	Mean	STDev
Level A	1413	12	9424	1393	1396	136.38	149.96
Level AA	1032	12	10,092	1408	1376	139.2	143.28
Level AAA	1136	12	9547	1413	1367	134.75	130.39
Mean	59.68	0.6	484.38	70.23	68.98		
STDev	67.57	1.92	450.96	112.2	73.37		

---

[8]https://wave.webaim.org/
[9]https://tenon.io/

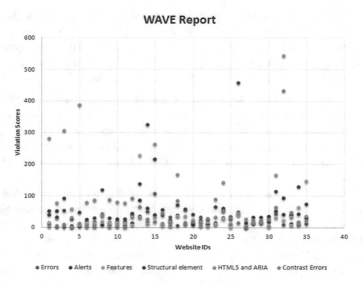

Fig. 5 A violation scores of selected corporation websites based on WAVE tool

Table 3 Detection of worst performing web pages among the 35 websites

Worst performing web pages (Tenon tool)				
S. No.	OTA website	Total issue	Error density	Warning density
5	https://www.homeaway.com/	1007	54	51
8	https://www.hotwire.com/	2155	66	168
13	https://www.cruiseshipcenters.com/	1826	183	24
14	https://www.classicvacations.com/	1437	133	50
15	https://www.tripadvisor.com/	3585	129	39
16	https://www.airfarewatchdog.com/	1869	116	88
18	https://www.familyvacationcritic.com/	818	56	43
21	https://www.holidaylettings.co.uk/	799	96	9
24	https://www.jetsetter.com/	916	89	27
26	https://www.seatguru.com/	5212	365	163
27	https://hotels.tingo.travel/	100	10	5
28	https://www.vacationrentals.com/	424	49	10
29	https://www.viator.com/	1269	94	47
30	https://www.onetime.com/	390	27	24
31	https://www.booking.com/	1451	133	31
34	https://www.agoda.com/	956	83	31
35	https://www.rentalcars.com/	817	67	32

Table 4 Tenon tool result of 35 websites: issue by content category

Tenon tool result: issue by content category		
Features	Total issues	Issue percentage
Images and other non-text content	3186	10.07%
Tables	222	0.70%
Cascading Style Sheets (CSS)	6155	19.47%
Forms	1094	3.46%
Navigation	2874	9.09%
Frames and i-frames	70	0.22%
Document structure	4733	14.97%
Language, typography, and content	10,953	34.64%
Dynamic content	0	0.00%
Multimedia	4	1.20%
Keyboard accessibility & focus control	1207	3.81%
Custom controls	1114	3.52%

Table 5 Readability analysis of 35 OTA websites based on three international corporations

Readability analysis						
Corporation websites	FKRE	FKGL	GFOG	SMOG	CLI	ARI
Readability index scores	69.9	5.25	6	5.6	10	2.9
Average grade level	6					

5.2 Easy to Read Result

The main motive of these selected websites is to communicate with the customers to share information and resources. At the same time, their easy to read status is an important factor for easy to access resources. Table 5 presents the readability scores of different indices[10] and the overall grade level of understandability of websites based on three international corporations. It is found that their average grade level of readability comes under 6 of US grade. It should be minimized further so that more easy to read status of websites can be achieved.

5.3 Ranking Relevance and Correlations

A relevance of ranking based on violation scores obtained from different techniques along with site visiting ranking of 35 OTA websites is shown in Fig. 6, using rank cases in SPSS procedure to convert violation scores into ranking to determine the importance of association between the accessibility, readability, and site rankings.

---

[10]https://www.webfx.com/tools/read-able/

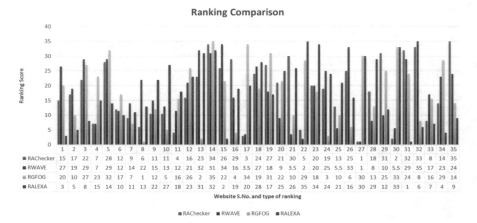

Fig. 6 A relevance of ranking based on violation scores obtained from different techniques along with site visiting ranking of 35 OTA websites

Table 6 Correlation between the variables AChecker, GFOG, WAVE, and Alexa

Correlations		AChecker	GFOG	WAVE	Alexa
AChecker	Pearson correlation	1.00	0.12	0.57	−0.15
	Sig. (2-tailed)		0.494	0.000	0.398
	N	35	35	35	35
GFOG	Pearson correlation	0.12	1.00	−0.08	0.26
	Sig. (2-tailed)	0.494		0.642	0.129
	N	35	35	35	35
WAVE	Pearson correlation	0.57	−0.08	1.00	−0.17
	Sig. (2-tailed)	0.000	0.642		0.340
	N	35	35	35	35
Alexa	Pearson correlation	−0.15	0.26	−0.17	1.00
	Sig. (2-tailed)	0.398	0.129	0.340	
	N	35	35	35	35

Table 6 presents the Pearson correlation between AChecker, GFOG, WAVE, and Alexa variables of accessibility, readability, and site ranking. It is found that the correlation between accessibility and readability is positively weak, and intra-accessibility correlation is positively strong. But, the correlation of accessibility with site ranking is negative, and readability with site ranking is positive. Thus, it is suggested that if the violations of accessibility guidelines are minimized, and the readability of web content is enhanced, it may result in more positive correlation among the websites.

Table 7 Statistical performance score cum relevance of attributes among the selected three international corporation websites

Statistical inference of 35 OTA websites

Corporation name	Website S. No.	AChecker		WAVE		GFOG	
		Mean	STDev	Mean	STDev	Mean	STDev
Expedia Group Inc.	1–14	1556.86	1360.60	280.64	163.57	5.42	2.74
TripAdvisor Inc.	15–30	1701.31	979.83	226	193.63	5.83	2.04
Booking Holdings Inc.	31–35	3244.8	2403.29	475.6	385.67	5.52	1.03

5.4   Statistical Analysis of Corporations

Table 7 presents the statistical performance score cum relevance of attributes among the selected three international corporations websites. It is found that there is a maximum accessibility violation in Booking Holdings Inc. than the other two. But the readability score is almost the same. In a nutshell, there is a need to focus the parameters of accessibility and readability to get rid of these issues so that the universality of these websites is achieved and accessible tourism may be enhanced.

6   Suggestions and Implications

Web accessibility and readability are an important aspect for the success and acceptance of OTA websites globally by the users especially PwDs. However, from the result it has been found that most of the websites failed to comply with one or more guidelines of WCAG 2.0, and the readability level is also quite high. This is the dominant issue in the field of tourism and hospitality that creates the barrier for PwDs to access the information over the Internet [25, 79, 91]. Rules and regulations are made to address these issues related to PwDs, but it is not strictly enforced. In some countries like the USA, the UK, and Australia, violating the standards of web accessibility is against the law of disability and would be subject to a lawsuit [45, 54]. These laws and regulations related to accessibility should be more extensive and include travel and tourism industry to strictly enforced accessibility as primary criteria.

It is not only the responsibility of the government to enforce the accessibility regulation, but it is also the responsibility of the OTAs to follow the accessibility guidelines, to make their websites accessible for accommodating the niche market segment. Bridging the gap between tourism and PwDs is not only important for

human right and freedom to promote accessible tourism and sustainable development [74] but also a business opportunity for OTAs to include the new customers in
their business, since they likely to participate in tourism at low season, they mostly
travel in groups, and in some region, these groups spend more than average on their
trips [77]. Accessibility issues of a website could be addressed at the beginning
because the cost is minimal at the beginning, and if it is not resolved at that time,
then the accumulated cost is higher, and it is termed as accessibility debt [78].
Therefore, OTAs should take care of accessibility issues from the beginning and
update it regularly according to guidelines. Accessibility is a continuous process
because of changes in the external environment and consumer behavior; hence,
OTAs are updating their websites often to accommodate their customers' needs and
expectations.

In addition to accessibility, readability is also a critical factor of a website for the
users including PwDs. Texts that are difficult to read and understand would create
barrier for the user while using the website. Hence, OTAs should design their content
of the website according to the understanding and readability expectations of their
users and avoid complex words to improve readability. It does not matter what the
website is portraying if their users cannot understand it. Indeed, PwDs are often
more loyal toward the website that respects their access needs [47]. The indices used
for determining web readability are based on the English language. Therefore, there
is a need of readability tools that should be independent of language. Moreover, the
readability was measured on US grade level for understanding the text [34]. Thus,
the researchers should focus to build readability tools that measure country-specific
grade level or tools that are universal in nature irrespective of the country grade
level.

In order to make a web page accessible as per the international or national
guidelines, it is important that web developer or web designer should be aware of
WCAG or any other standard guidelines. However, previous research reported that
they do not have knowledge on WCAG and are not familiar with the standards and
technologies used by PwDs [31]. The limited awareness and knowledge and absence
of training for personnel who are responsible for the accessibility of the website like
content creators and developers can bring about barriers in making the web page
accessible [5]. Therefore, it is necessary for the organizations and government to
provide adequate training on web accessibility to minimize complexity universally,
and training should be based on the latest guidelines of web accessibility.

7   Conclusion

In the twenty-first century, web accessibility and readability issues are the most
discussed and researched subjects of the WWW due to their significant impact on
people with access needs and to provide equal participation of PwDs in Information
and Communication Technologies (ICT). However, the issues of web accessibility
and readability in the travel and tourism field are yet to unearth from WCAG 2.0

and readability index perspectives. To address these issues, this study has analyzed that accessibility and readability along with performance and site visiting ranking of 35 OTA websites belong to three international corporations.

Based on the result obtained, it has been found that websites (35 OTAs) are not successfully implementing the WCAG guidelines and failed in one or more guidelines of WCAG 2.0. The most violated issues such as *language, content and typography, CSS, document structure, and image and other non-text content* should be minimized to improve accessibility. In addition to this, warnings cum alerts should be minimized, and accessibility component features should be added to OTA websites for further enhancement. Also, it has been found that their average grade of readability is 6, which needs to be minimized further so that readability of web content is attained. Therefore, OTAs should simplify their text to improve their performance on readability so that users having lower-grade level could have a better (more easy) understanding of the text.

Therefore, it is suggested to web administrators and developers to remove the identified issues and make corresponding improvements in OTA websites regarding accessibility and readability standards that can improve their global ranking too. Doing this, correlations between the said variables may be positively strengthened.

In the future, OTA websites need to take more inclusive steps to embrace PwDs as online users by reexamining, reevaluating, and reviewing their websites to ensure swift navigation inside the web page and overall accessibility cum readability.

References

1. AChecker. (2019). Web accessibility checker. Available at: https://achecker.ca/checker/index.php. Accessed 15 Apr 2019.
2. Adam, A., & Kreps, D. (2009). Disability and discourses of web accessibility. *Information, Communication & Society, 12*(7), 1041–1058.
3. Alayed, A., Wald, M., & Draffan, E. A. (2016). A framework for the development of localised web accessibility guidelines for university websites in Saudi Arabia. In M. Antona & C. Stephanidis (Eds.), *Universal access in human-computer interaction* (Methods, techniques, and best practices, pp. 3–13). Cham: Springer International Publishing.
4. Alén, E., Domínguez, T., & Losada, N. (2012). New opportunities for the tourism market: Senior tourism and accessible tourism. In *Visions for global tourism industry-creating and sustaining competitive strategies*. Rijeka: IntechOpen.
5. Ballesteros, E., Ribera, M., Pascual, A., & Granollers, T. (2015). Reflections and proposals to improve the efficiency of accessibility efforts. *Universal Access in the Information Society, 14*(4), 583–586.
6. Bi, Y., Card, J. A., & Cole, S. T. (2007). Accessibility and attitudinal barriers encountered by Chinese travellers with physical disabilities. *International Journal of Tourism Research, 9*(3), 205–216.
7. Bisht, P. (2019). Online travel market report. Available at: https://www.alliedmarketresearch.com/online-travel-market. Accessed 12 May 2019.
8. Bowtell, J. (2015). Assessing the value and market attractiveness of the accessible tourism industry in Europe: A focus on major travel and leisure companies. *Journal of Tourism Futures, 1*(3), 203–222.

9. Buhalis, D. (1998). Information technologies in tourism: Implications for the tourism curriculum. In *Information and Communication Technologies in Tourism 1998* (pp. 289–297). Wien: Springer.
10. Buhalis, D. (2003). eTourism: Information technology for strategic tourism management. London, UK: Pearson Education.
11. Buhalis, D., & Law, R. (2008). Progress in information technology and tourism management: 20 years on and 10 years after the internet—The state of eTourism research. *Tourism Management, 29*(4), 609–623.
12. Buhalis, D., & Michopoulou, E. (2011). Information-enabled tourism destination marketing: Addressing the accessibility market. *Current Issues in Tourism, 14*(2), 145–168.
13. Buhalis, D., & O'Connor, P. (2005). Information communication technology revolutionizing tourism. *Tourism Recreation Research, 30*(3), 7–16.
14. Buhalis, D., Eichhorn, V., Michopoulou, E., & Miller, G. (2005). *Accessibility market and stakeholder analysis.* University of Surrey y One Stop Shop for Accessible Tourism in Europe (OSSATE).
15. Burger, F., Kroiß, P., Pröll, B., Richtsfeld, R., Sighart, H., & Starck, H. (1997). Tis@ web-database supported tourist information on the web. In *Information and Communication Technologies in Tourism 1997* (pp. 180–189). Wien/New York: Springer.
16. Connolly, D. J., Olsen, M. D., & Moore, R. G. (1998). The internet as a distribution channel. *Cornell Hotel and Restaurant Administration Quarterly, 39*(4), 42–54.
17. Dale, C. (2003). The competitive networks of tourism e-mediaries: New strategies, new advantages. *Journal of Vacation Marketing, 9*(2), 109–118.
18. Darcy, S. (2010). Inherent complexity: Disability, accessible tourism and accommodation information preferences. *Tourism Management, 31*(6), 816–826.
19. Darcy, S., Cameron, B., Pegg, S., et al. (2011). Developing a business case for accessible tourism. In *Accessible tourism: Concepts and issues* (pp. 241–259). Bristol/Toronto: Channel View Publications.
20. Disability Rights Commission. (2004). *The web: Access and inclusion for disabled people; a formal investigation.* London: The Stationery Office.
21. Domínguez Vila, T., Alén González, E., & Darcy, S. (2018). Website accessibility in the tourism industry: An analysis of official national tourism organization websites around the world. *Disability and Rehabilitation, 40*(24), 2895–2906.
22. Domínguez Vila, T., Alén González, E., & Darcy, S. (2019). Accessible tourism online resources: A northern European perspective. *Scandinavian Journal of Hospitality and Tourism, 19*(2), 140–156.
23. DuBay, W. H. (2004). *The principles of readability.* Online Submission.
24. Eichhorn, V., Miller, G., Michopoulou, E., & Buhalis, D. (2008). Enabling access to tourism through information schemes? *Annals of Tourism Research, 35*(1), 189–210.
25. ENAT. (2013). Accessibility review of European national tourist boards' websites 2012. Available at: https://www.accessibletourism.org/resources/enat-nto-websites-study-2012_public.pdf. Accessed 11 May 2019.
26. Gonçalves, R., Martins, J., Pereira, J., Oliveira, M. A. Y., & Ferreira, J. J. P. (2013). Enterprise web accessibility levels amongst the forbes 250: Where art thou o virtuous leader? *Journal of Business Ethics, 113*(2), 363–375.
27. Gutierrez, C. F., Loucopoulos, C., & Reinsch, R. W. (2005). Disability-accessibility of airlines' web sites for us reservations online. *Journal of Air Transport Management, 11*(4), 239–247.
28. Hanna, J., & Millar, R. (1997). Promoting tourism on the internet. *Tourism Management, 18*(7), 469–470.
29. Ho, C. I., & Lee, Y. L. (2007). The development of an e-travel service quality scale. *Tourism Management, 28*(6), 1434–1449.
30. Hyun, J., Moon, J., & Hong, K. (2008). Longitudinal study on web accessibility compliance of government websites in Korea. In S. Lee, H. Choo, S. Ha, & I. C. Shin (Eds.), *Computer-human interaction* (pp. 396–404). Berlin/Heidelberg: Springer.

31. Inal, Y., Rızvanoğlu, K., & Yesilada, Y. (2017). Web accessibility in Turkey: Awareness, understanding and practices of user experience professionals. *Universal Access in the Information Society, 18*, 1–12.
32. Ismail, A., & Kuppusamy, K. S. (2016). Accessibility of Indian universities' homepages: An exploratory study. *Journal of King Saud University-Computer and Information Sciences*, 268–278.
33. Ismail, A., & Kuppusamy, K. S. (2019). Web accessibility investigation and identification of major issues of higher education websites with statistical measures: A case study of college websites. *Journal of King Saud University-Computer and Information Sciences*, in press.
34. Ismail, A., Kuppusamy, K. S., Kumar, A., & Ojha, P. K. (2017). Connect the dots: Accessibility, readability and site ranking-an investigation with reference to top ranked websites of government of india. *Journal of King Saud University-Computer and Information Sciences*, 528–540.
35. Ismail, A., Kuppusamy, K. S., & Nengroo, A. S. (2018). Multi-tool accessibility assessment of government department websites: A case-study with JKGAD. *Disability and Rehabilitation: Assistive Technology, 13*(6), 504–516.
36. Ismail, A., Kuppusamy, K. S., & Paiva, S. (2019). Accessibility analysis of higher education institution websites of Portugal. *Universal Access in the Information Society*, 1–16.
37. Ismailova, R., & Kimsanova, G. (2017). Universities of the Kyrgyz republic on the web: Accessibility and usability. *Universal Access in the Information Society, 16*(4), 1017–1025. https://doi.org/10.1007/s10209-016-0481-0.
38. Jeong, M., & Lambert, C. (1999). Measuring the information quality on lodging web sites. *International Journal of Hospitality Information Technology, 1*(1), 63–75.
39. Kaynama, S. A., & Black, C. I. (2000). A proposal to assess the service quality of online travel agencies: An exploratory study. *Journal of Professional Services Marketing, 21*(1), 63–88.
40. Kingman, A. (2018). A brief history of WCAG. Available at: https://lastcallmedia.com/blog/brief-history-wcag. Accessed 17 May 2019.
41. Law, R., & Hsu, C. H. (2005). Customers' perceptions on the importance of hotel web site dimensions and attributes. *International Journal of Contemporary Hospitality Management, 17*(6), 493–503.
42. Law, R., Qi, S., & Buhalis, D. (2010). Progress in tourism management: A review of website evaluation in tourism research. *Tourism Management, 31*(3), 297–313.
43. Lukaitis, A., & Davey, B. (2012). Web design for mature-aged travellers: Readability as a design issue. *Journal of Marketing Development and Competitiveness, 6*(2), 69–80.
44. Maswera, T., Dawson, R., & Edwards, J. (2005). Analysis of usability and accessibility errors of e-commerce websites of tourist organisations in four African countries. In *Information and Communication Technologies in Tourism 2005* (pp. 531–542).
45. Maurer, R. (2018). Number of federal website accessibility lawsuits nearly triple, exceeding 2250 in 2018.
46. Mills, J. E., Han, J. H., & Clay, J. M. (2008). Accessibility of hospitality and tourism websites: A challenge for visually impaired persons. *Cornell Hospitality Quarterly, 49*(1), 28–41.
47. Nielsen, J. (1999). *Designing web usability: The practice of simplicity*. Berkeley: New Riders Publishing.
48. O'Connor, P., & Frew, A. J. (2004). An evaluation methodology for hotel electronic channels of distribution. *International Journal of Hospitality management, 23*(2), 179–199.
49. O'Connor, P. (2004). Privacy and the online travel customer: An analysis of privacy policy content, use and compliance by online travel agencies (A. J. Frew, ed.). New York: Springer-Verlag New York.
50. Oertel, B., Hasse, C., Scheermesser, M., Thio, S. L., & Feil, T. (2004). Accessibility of tourism web sites within the European Union. In *Proceedings of the 11th international conference on information and communication Technologies in Tourism (ENTER 2004)* (pp. 358–368). Cairo: Springer.

51. Ojha, P. K., Ismail, A., & Kuppusamy, K. S. (2018). Perusal of readability with focus on web content understandability. *Journal of King Saud University-Computer and Information Sciences*, in press.
52. O'reilly, T. (2007). What is web 2.0: Design patterns and business models for the next generation of software. *Communications & strategies, 65*(1), 17.
53. O'reilly, T. (2009). *What is web 2.0*. Massachusetts, USA: O'Reilly Media, Inc.
54. Outlaw. (2011). *Disabled access to websites under UK law*. Available at: https://www.out-law.com/page-330. Accessed 01 May 2019.
55. Paciello, M. (2000). *Web accessibility for people with disabilities*. Florida, USA: CRC Press.
56. Parasuraman, A., Zeithaml, V. A., & Malhotra, A. (2005). ES-QUAL: A multiple-item scale for assessing electronic service quality. *Journal of Service Research, 7*(3), 213–233.
57. Pashkevich, A., & Stjernström, O. (2014). Making Russian arctic accessible for tourists: Analysis of the institutional barriers. *Polar Geography, 37*(2), 137–156.
58. Philstar. (2019). 5 ways online travel agencies make traveling a lot easier. Available at: https://www.philstar.com/lifestyle/travel-and-tourism/2016/12/30/1658089/5-ways-online-travel-agencies-make-traveling-lot-easier. Accessed 10 May 2019.
59. Poon, A., et al. (1993). *Tourism, technology and competitive strategies*. Wallingford: CAB International.
60. Ritchie, H., & Blanck, P. (2003). The promise of the internet for disability: A study of on-line services and web site accessibility at centers for independent living. *Behavioral Sciences & the Law, 21*(1), 5–26.
61. Schegg, R., Liebrich, A., Scaglione, M., & Ahmad, S. F. S. (2008). An exploratory field study of web 2.0 in tourism. In *Information and Communication Technologies in Tourism 2008* (pp. 152–163).
62. Sheldon, P. J., et al. (1997). *Tourism information technology*. Wallingford: Cab International.
63. Shi, Y. (2006). The accessibility of queensland visitor information centres' websites. *Tourism Management, 27*(5), 829–841.
64. Shneiderman, B., & Plaisant, C. (2010). *Designing the user interface: Strategies for effective human-computer interaction*. Massachusetts, USA: Pearson Education India.
65. Sigala, M. (2007). *Web 2.0 in the tourism industry: A new tourism generation and new e-business models*. Greece: Travel Daily News.
66. Sigala, M. (2012). Web 2.0 and customer involvement in new service development: A framework, cases and implications in tourism. In *Web* (Vol. 2, pp. 25–38).
67. Smith, R. W. (1987). Leisure of disable tourists: Barriers to participation. *Annals of Tourism Research, 14*(3), 376–389.
68. Smith, C., Jenner, P., et al. (1998). Tourism and the internet. *Travel & Tourism Analyst, 1*(1), 62–81.
69. Staszak, J. (2018). Three stock experts take on the online travel world. Available at: https://www.forbes.com/sites/moneyshow/2018/03/16/three-stock-experts-take-on-the-online-travel-world/#4b1210b796ab. Accessed 01 May 2019.
70. Sussmann, S., & Baker, M. (1996). Responding to the electronic marketplace: Lessons from destination management systems. *International Journal of Hospitality Management, 15*(2), 99–112.
71. Tjoa, A. M., & Werthner, H. (1996). Interfacing WWW with distributed database applications in the field of tourism. In *Information and Communication Technologies in Tourism* (pp. 78–85). Wien: Springer.
72. Toms, E. G., & Taves, A. R. (2004). Measuring user perceptions of web site reputation. *Information Processing & Management, 40*(2), 291–317.
73. UN. (2006). United nations convention on the rights of persons with disabilities. Available at: https://www.un.org/disabilities/documents/convention/convention_accessible_pdf.pdf. Accessed 13 May 2019.
74. UNWTO. (2013). Recommendations on accessible tourism. Available at: http://cf.cdn.unwto.org/sites/all/files/pdf/unwto_recommendations_on_accessible_tourism.pdf. Accessed 17 May 2019.

75. UNWTO. (2016). Manual on accessible tourism for all: Principles, tools and best practices. Available at: https://www.e-unwto.org/doi/pdf/10.18111/9789284418077. Accessed 14 May 2019.
76. UNWTO. (2016). Manual on accessible tourism for all: Principles, tools and best practices. Available at: http://cf.cdn.unwto.org/sites/all/files/docpdf/moduleieng13022017.pdf. Accessed 10 May 2019.
77. UNWTO. (2016). Tourism for all – promoting universal accessibility. Available at: https://www.e-unwto.org/doi/pdf/10.18111/9789284418138. Accessed 17 May 2019.
78. Vera, C. L. (2018). The true cost of universal accessibility. Available at: https://uxdesign.cc/the-true-cost-of-universal-accessibility-7e496d678a9f. Accessed 17 May 2019.
79. Vila, T. D., González, E. A., & Darcy, S. (2019). Accessibility of tourism websites: The level of countries' commitment. In *Universal Access in the Information Society* (pp. 1–16).
80. W3C. (1999). Web content accessibility guidelines 1.0. Available at: https://www.w3.org/TR/1999/WAI-WEBCONTENT-19990505/wai-pageauth.html. Accessed 17 May 2019.
81. W3C. (2008). Web content accessibility guidelines (wcag) 2.0. Available at: https://www.w3.org/TR/WCAG20/. Accessed 01 May 2019.
82. W3C. (2018). Web content accessibility guidelines (wcag) 2.1. Available at: https://www.w3.org/TR/WCAG21/. Accessed 01 May 2019.
83. W3C. (2019). Introduction to web accessibility. Available at: https://www.w3.org/WAI/fundamentals/accessibility-intro/#context. Accessed 10 May 2019.
84. Waldron, V. R., Lavitt, M., & Kelley, D. (2000). The nature and prevention of harm in technology-mediated self-help settings: Three exemplars. *Journal of Technology in Human Services, 17*(2–3), 267–293.
85. Wang, Y. S., & Tang, T. I. (2004). Assessing customer perceptions of web site service quality in digital marketing environments. In *Advanced topics in end user computing* (Vol. 3, pp. 16–35). Hershey: IGI Global.
86. Watch, M. (2019). Online travel agency (OTA) market to witness huge growth by 2025. Available at: https://www.marketwatch.com/press-release/online-travel-agency-ota-market-to-witness-huge-growth-by-2025-booking-holdings-tripadvisor-expedia-homeaway-makemytrip-kayak-qunr-2019-03-09. Accessed 09 May 2019.
87. Wen, I. (2009). Factors affecting the online travel buying decision: A review. *International Journal of Contemporary Hospitality Management, 21*(6), 752–765.
88. Williams, R., & Rattray, R. (2005). UK hotel web page accessibility for disabled and challenged users. *Tourism and Hospitality Research, 5*(3), 255–268.
89. Williams, R., Rattray, R., & Stork, A. (2004). Web site accessibility of German and UK tourism information sites. *European Business Review, 16*(6), 577–589.
90. Williams, R., Rattray, R., & Grimes, A. (2006). Meeting the on-line needs of disabled tourists: An assessment of UK-based hotel websites. *International Journal of Tourism Research, 8*(1), 59–73.
91. Williams, R., Rattray, R., & Grimes, A. (2007). Online accessibility and information needs of disabled tourists: A three country hotel sector analysis. *Journal of Electronic Commerce Research, 8*(2), 157–171.
92. World Health Organization, et al. (2011). *World report on disability 2011*. Geneva: World Health Organization.
93. Xiong, L., Cobanoglu, C., Cummings, P., & DeMicco, F. (2009). Website accessibility of US based hospitality websites. In *Information and Communication Technologies in Tourism 2009* (pp. 273–284).
94. Yoo, B., & Donthu, N. (2001). Developing a scale to measure the perceived quality of an internet shopping site (sitequal). *Quarterly Journal of Electronic Commerce, 2*(1), 31–45.

How Artificial Intelligence Can Undermine Security: An Overview of the Intellectual Property Rights and Legal Problems Involved

Praveen Kumar Gupta, Deepali Venkatesh Prasanna, and Shreeya Sai Raghunath

1 Introduction

Artificial Intelligence (AI), also commonly called machine intelligence, is a niche branch of computer science that deals with the computation and interpretation of complex data. It has the incredible ability to draw useful information from within a data set to reach conclusions in the fields of but not related to – Diagnosis, Treatment, and Outcome prediction in several clinical applications. The advent of Artificial Intelligence or machine learning has greatly reduced time and increased the amount of data that can be processed. Let us take a look at the different types of Artificial Intelligence and how they work, briefly (Fig. 1).

Artificial Neural Network This model bases itself on a collection of connected nodes, also referred to as artificial neurons, which are largely modeled after the neurons of the human brain. Each node acts as synapses of the biological brain and has the ability to transmit a signal from one artificial neuron to another. The artificial neuron can process the signal it receives. The signal received at the junction between each artificial neuron is a real number, and its output is calculated by a nonlinear function of the summation of its inputs. The connections are called "edges." Each artificial neuron and its edge has a weight that alters as the learning proceeds. The weight can either intensify or decrease the strength of the signal at each edge. Artificial neurons generally have a threshold as a result of which, the signal will only be sent if the sum of the signals crosses that threshold. The artificial neurons are aggregated into layers which may perform different kinds of transformations

P. K. Gupta (✉) · D. V. Prasanna · S. S. Raghunath
Department of Biotechnology, R.V. College of Engineering, Bangalore, India
e-mail: praveenkgupta@rvce.edu.in

© Springer Nature Switzerland AG 2021
R. Kumar, S. Paiva (eds.), *Applications in Ubiquitous Computing*, EAI/Springer Innovations in Communication and Computing,
https://doi.org/10.1007/978-3-030-35280-6_3

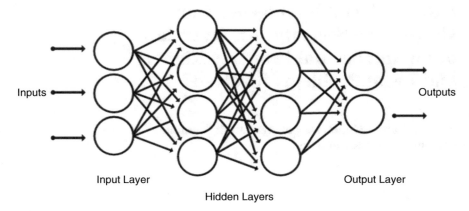

Fig. 1 Neural networks, which are organized in layers consisting of a set of interconnected nodes. Networks can have tens or hundreds of hidden layers [1]

upon their inputs. Signals move from the first layer (called the input layer), to the last layer (called the output layer), after crossing the layers several times [2].

Search and Optimization Problems in AI can be resolved theoretically by performing intelligent searches of the potential solution. For example, logical proof can be visualized as looking for a path that leads from origin to results, wherein each step is essentially the application of an interference rule. Planning algorithms sift through several trees of goals and subgoals, in an attempt to find a path to the desired target goal. This process is called means to an end analysis. Many learning algorithms use search algorithms based on techniques of optimization. Simple exhaustive searches are barely enough for most real-world problems: the search space rapidly grows to huge numbers of the astronomical order. The result is a slow or incomplete search. The solution is therefore to use Heuristics or "rules of thumb" that ranks choices in order of those more likely to reach the goal in a shorter number of steps. In some search procedures, heuristics can be used to entirely remove choices that are unlikely to lead to the goal. This is called pruning [3].

Logics It is utilized for representation of knowledge and problem-solving; however, it can be applied to other problems as well. Several different forms of logic are used in AI research.

Propositional Logic involves the use of Truth Functions like "or" and "not."
First-order logic sum quantifiers and predicates and thus displays facts about objects, their characteristics, and their interrelations.
Fuzzy Set Theory designates a "degree of truth" ranging from 0 to 1, to vague statements that are too loose to be considered definitively true or false. Fuzzy Logic has been successfully implemented in control systems. Fuzzy logic however does not do well in knowledge bases. Default logics, nonmonotonic logic are various types of logics that have been designed to aid with default reasoning and the qualification problem [3].

Deep Learning is an Artificial Neural Network that has the ability to learn from a long chain of causal links. For example, a feedforward network having six hidden layers can study a 7-link chain of six hidden layers and an output layer and has a Credit Assignment Path (CAP) depth of seven. Many deep learning systems require chains of ten or more causal links in length to learn. Deep learning has changed several pertinent subfields of Artificial Intelligence, including computer vision, Natural language processing, speech recognition, etc. Deep learning is based on Convolution Neural Networks. Simply put, Deep learning enables computers to learn in a manner similar to the way humans learn, that is, by example. The concept of Deep Learning was theorized in the early 1980s. However, it gained importance due to two main reasons:

(a) Deep learning utilizes huge amounts of *labeled data*. For example, for the development of driverless cars, it needs millions of images and hundred thousand of hours of video.
(b) Deep learning needs considerable *computing power*. High-performance graphics processing units (GPUs) contain a parallel architecture that is very useful for deep learning. When this is combined with clusters or cloud-based computing, it allows development teams to decrease the training duration for a deep learning network from several months to a few short hours or perhaps even less [4].

Deep learning models are trained by utilization of huge sets of labeled data and neural network architectures that directly learn features from the data set without manual feature extraction.

Artificial Intelligence helps to improve the credibility of the data available and helps to improve the patient's privacy when the data are stored on the cloud. However, improper and/or excessive use of Artificial Intelligence can lead to unintended consequences. In this chapter, we will be focusing on the general applications of Artificial Intelligence in the various fields of Biomedical engineering, the various issues faced with and without Artificial Intelligence, and possible solutions. In the following sections, we will read and understand about the various applications of Artificial Intelligence in the healthcare field, followed by sections that will shed light upon what are the various problems or threats associated with the usage of Artificial Intelligence. The major part of this chapter however will be focused on understanding the numerous solutions to problems.

2 Applications of Artificial Intelligence

With the surplus of medical data that is being collected by physicians and other medical professionals, there is a single question that dawns on the mind: What to do with these data? With the incorporation of Artificial Intelligence in the medical industry, the time and effort that was previously invested by doctors and physicians to make sense of the immense data has greatly reduced the time invested by them to analyze and make useful interpretations of the data. Apart from reducing time, the use of Artificial Intelligence has also greatly reduced the need for extra and skilled labor as the same work can now be done using a single computer system. With the use of electronic or digitized medical records, doctors can keep a track of the patient's entire medical history and clearly identify the various drugs or treatments that a patient has been given, in chronological order. Further, autogenerated messages, alarms, reminders, and alerts can be sent to the patient to inform them of an upcoming appointment or convey other information. The same can be also utilized to send reminders to doctors to carry out tests for the patient on their next visit.

Some of the most widely used applications of Artificial Intelligence in the Medical Industry include (Figs. 2 and 3):

Clinical Decision Support System A Clinical Decision Support System (CDSS) is a system that helps doctors, nurses, and physicians to make informed decisions. It helps to increase their effectiveness and efficiency by providing prompts or advice during the decision-making process. This is incorporated within the electronic health record. While the doctor is filling up the record after a checkup, the CDSS provides prompts to the user regarding diagnosis, treatment, possible drug interactions, allergic reactions, new drugs in the market, and other relevant information. The main goal is to increase the awareness and effectiveness of patient diagnosis

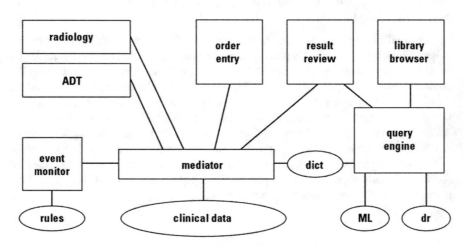

Figure. Architecture of the Integrated Advanced Information Management System.
ADT   = admission discharge transfer (administrative information system)
event monitor   = monitors messages passed through the system for opportunities to invoke pre-determined rules
clinical data   = large research and quality assurance database containing clinical information
dict   = standardized dictionary of terms used in the system
ML   = Medline for automated literature searches
dr   = drug information knowledge base

Reproduced, with permission, from Hripcsak GH, IAIMS architecture, J Am Med Inform Assoc 1997;4:S20-S30.

Fig. 2 Clinical decision support systems [5]

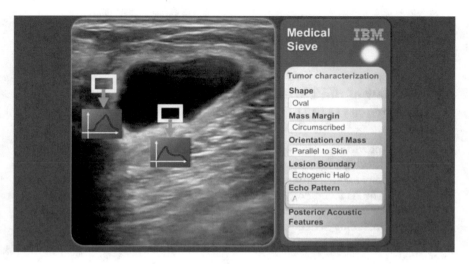

Fig. 3 IBM Watson Health care and its application in radiology [6]

and improve the overall health outcome. It also links the access to large amounts of literature, so that the user can, at any time, look up the material to make informed decisions [7].

Telemedicine Telemedicine is the use of technology, more specifically telecommunications and information technology, to remotely monitor patients. This has proved to be revolutionary in the treatment of patients at distant inaccessible areas. This has also greatly reduced the time and money spent in healthcare and has helped people with the white coat syndrome. Additionally, patients suffering from multiple sclerosis, Parkinson's, and other diseases that cause immobility no longer need to travel for frequent checkups. Renewal and verification of prescription can also take place online thanks to telemedicine [8].

Radiology The field that has gained the maximum attention is the field of Radiology. The ability of a computer to interpret imaging results can possibly help clinicians in detecting change in the radio film image that a clinician may have accidentally missed. Studies at Stanford led to the creation of an algorithm that could detect pneumonia at a specific site, in the test patients, with a better average F1 metric than the radiologists participating in the trial. The Radiological Society of North America has had several presentations on AI in imaging at its annual meeting. The emergence of AI technology in radiology is perceived as a threat by some specialists, as the technology can achieve improvements in certain statistical metrics in isolated cases, as opposed to specialists [9].

Drug Development Artificial Intelligence is being increasingly utilized along with bioinformatics tools to help in the in silico drug discovery process. It can be used to analyze large amounts of molecular information to search for potential drug compounds and can be used for target analysis. This could potentially reduce the overall time taken in the drug discovery and testing process and save millions [10].

3 Overview of the Threats Associated with Artificial Intelligence [11, 12]

As discussed previously, Artificial Intelligence systems are a promising technology capable of mimicking the human learning process. They can even execute certain tasks without any explicit programming and make their own decisions based on past experiences. The AI technology already has a home in the Information Technology (IT) sector and as with all technology that uses data and networks, it is subject to malpractices including hacking, data theft, and reprogramming.

There are three main areas, in which the security of Artificial Intelligence is threatened:

- Physical Security: AI systems being employed in controlling physical devices such as drones can be hacked and misused by cyber criminals. These perpetrators may orchestrate complicated and covert attacks on the system that may be difficult to identify and remedy and have the potential to compromise security at a global level.
- Political Security: AI technologies can be used by hackers to analyze mass collected data and manipulate surveillance systems. These techniques can be used to target a specific political group.
- Digital Security: Manipulation by AI networks can make cyber threats such as viruses even more dangerous due to conferred adaptive and predictive nature. It can be used in the creation of new viruses and worms based on the existing ones and spread such threats across a network much more rapidly.

One of the most daunting aspects of hacking in Artificial Intelligence systems is that due to their ability to learn, once hacked, they do not require user control to fulfill malicious intent. The system is capable of making its own decisions based on past learning experiences. Using their expertise, the system can create much more sophisticated and covert malware that can be continuously improved. This property makes it difficult to remedy the malware using normal human intelligence and speed.

Another issue is that AI systems normally function in conjunction with enormous stores of sensitive data, so the threat of data leakage and manipulation that exists in other technologies such as Electronic Health Records and Health Information Systems exists in these systems as well. Hackers can obtain access to important databases and alter data mining. This is an imperative security issue on the global scale with respect to industries such as the finance market, but not as consequential at the individual level.

Despite these security measures, many defense measures exist. These include penetration testing, identifying vulnerabilities (to prevent serious harm), etc. Ironically, Artificial Intelligence solution systems are the most ideal solution to cyber threats. The intention is to create self-learning solution generating systems. These AIs can monitor the existing threats and remedies and learn to format new remedies and provide security suggestions. Details of such systems are provided under the "Solutions" topic of this chapter.

AI defenses, however, do not come at a small cost. Fortunately, they develop enough interest to earn investment from companies and organizations. For example, $2 billion was earmarked by the Pentagon to help fund the Defense Advanced Research Project Agency. In the medical imaging industry, 130 million dollars of investment was raised by AI companies. Although it has been speculated that AI could eliminate the need for many human jobs, many recent studies have predicted an increase in Gross Domestic Product (GDP) and the generation of a new variety of jobs in place of the old ones.

Despite all this, it is still highly inaccurate to say that AI is a cure-all. Many pitfalls still exist in intelligent systems and much research and work is to be done to overcome these. Security has always been a prime concern to the public in terms of liberal utilization. Many people are suspicious or wary of such emerging technology, especially with respect to how far humans can control the intelligence conferred to a machine and with respect to the privacy and security of their data. Only once these concerns are addressed can AI tools be implemented on a much wider scale commercially. This is especially true in the case of AI used in the field of medicine. Medical records have a plethora of sensitive information and are very sensitive to malpractice. This is even truer considering that many medical records and cloud-stored information from wearable health monitoring systems contain nonmedically relevant information such as geographical location and personal details. However, in the event of successfully ensuring the safety and security of such devices the benefits would far outweigh the risks in these systems, changing the course and speed of progress in medicine.

The important problems and solutions regarding security in AI systems have been discussed in detail in the following sections:

4 Implications of Using Artificial Intelligence

The incorporation of AI-based technology into healthcare which has several benefits is expected to introduce a plethora of new risks and associated problems and intensify existing ones. For example, a simple failure in any commonly used software has the ability to rapidly and suddenly affect large groups of patients. Assumptions in underlying data and models, which may be hidden, can result in the AI-based systems suggesting dangerous recommendations that are largely inappropriate and not in lieu with the local care processes. Opaque techniques like deep learning can make explaining and learning from failure tremendously difficult.

To exploit the benefits of AI in healthcare and to build an honest environment for patients and practitioners, it is essential to robustly govern the risks that AI poses to patient safety. Given today's trend and us relying increasingly on the use of computers and Artificial Intelligence, it is possible that we will soon encounter problems that we did not expect and thus are not prepared to face. Some of the problems associated with using AI are – [13]:

Data Set The results of an Artificial Intelligence system are based on the data that are used to train it. Accuracy of the data plays an important role. While the AI system is modeled after the human brain, it does not have the ability to distinguish and make judgments for itself. It can only do what it is taught to do. As such, the data that are fed to it have to be as perfect as possible to obtain an accurate result. Feeding imperfect data can lead the system to produce false outcomes. For example, if patients suffering from pneumonia are said to be at high risk and moved to the intensive care unit (ICU) for treatment, patients suffering from asthma are said to be at low risk of death. Then, a patient suffering from both will be considered to be at high risk by a doctor and shifted to intensive care immediately. However, an imperfect data set can lead the system to declare the patient at a low-risk range and this could lead to the death of a patient, due to false diagnosis [13].

Human Contact Increased usage of AI can change the way in which patients interact with their doctors or physicians. Reduction in the physical checkup time leads to error in final diagnosis and reduces the communication between the patient and the caregiver. Often, it is a person who can offer more support to another than a computer [14].

Privacy Storing large amounts of data on the cloud for further processing by the artificial intelligent system makes it vulnerable to attack by cyber criminals such as hackers. Medical data, especially those obtained from wearable devices and from medical health records, contain highly sensitive information. This information can include geographical locations, routine activities, pictures, medical history, name, age, address, race, etc. If in the wrong hands, this has the potential to harm one's life. As such, protecting the privacy of the patient is a major issue. Data utilized for training the AI models cannot be used as a whole due to comprising the privacy [14].

Lack of Updated Facilities and Security Technology is advancing at a phenomenal rate. Hospitals need to update their computer systems and their security systems to keep up with the changing trends. However, this is not the case. Due to outdated systems, huge chunks of data are being accessed easily by online predators and unauthorized personnel.

While Artificial Intelligence has numerous benefits, there are a large number of ways in which its implementation can be improved, particularly in the security sector. The following section focuses on the solutions to the problems faced.

5 Solutions

This section involves a detailed description of the various types of solutions implemented with respect to the point of implementation. The areas covered include supporting evolving cloud architecture, protecting integrated medical devices, enabling the secure use of shared machines and workspaces, controlling identity management and databases, remote user authentication, remote device authentication, and Artificial Intelligence based solutions, along with various subtopics under each of them.

5.1 Supporting Evolving Cloud Architecture

The National Institute of Standards and Technology (NIST) defines cloud computing as "*a model for enabling ubiquitous, convenient, on-demand network access to a shared pool of configurable computing resources (e.g., networks, servers, storage, applications and services) that can be rapidly provisioned and released with minimal management effort and or service provider interaction*". Despite the numerous advantages of Cloud Computing, there are also multiple challenges among which data security is of prominence. The issues that are classified under data security are depicted in Fig. 4.

Solutions to Security Challenges in CIA, that is, Confidentiality, Integrity, and Availability, Include the Following:

- Data integrity can be checked for using Third Party Auditing (TPA).
- Data encryption is applied when the data are at rest and when they are in transit. Strong encryption algorithms like Rivest–Shamir–Adleman (RSA) and Advanced Encryption Standard (AES) algorithms are applied.
- Identity and Access Management (IAM) techniques must be implemented for user access to data.
- Intentional and accidental data changes can be found using the Hash Method. However, they are more time-consuming and consume more bandwidth.
- Encrypted data must not be stored along with the encryption key.

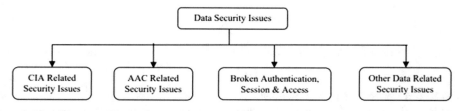

Fig. 4 The various issues that constitute data security issues [2]

- The availability issue can be mitigated by using the data dispersion technique, if other methods prove ineffective.
- Data redundancy, duplication, backups, and resilient systems are used to address issues regarding availability [15].

Solutions to Security Challenges in the Authentication and Access Control (AAC):
- Open standards, including Security Assertion Markup Language (SAML), must be applied for cloud computing., wherever applicable.
- Multifactor authentication enables both access and identity management and thus can be utilized.
- Biometric authentication, due to its uniqueness, is currently one of the most secure forms of single-sign-on authentication.
- Multiple third-party identity management solutions, such as Microsoft Azure Active Directory, Okta Identity Management, McAfee Cloud Identity Manager, etc., exist in the market.
- Single-sign-on policy must be adopted wherever it is possible to include such a feature.
- RSA cryptosystem can accept a variety of authentication models like knowledge-based authentication, two-factor authentication, and adaptive authentication [15].

Solutions to Security Challenges Due to Broken Authentication, Session, and Access Controls:
- A direct reference from an untrustworthy source should be checked for access control in order to make sure that the user is authorized for the requested resource.
- Automation must be applied to verify the deployment of proper authentication.
- Strong session management controls must be implemented.
- Cross-site Scripting (XSS) flaws must be avoided, as they can be used for session ID theft [15].

Solutions to Other Data-Related Security Issues:
- Data leakage should be avoided using strong encryption techniques for the data backup.
- Intelligent segregation techniques for data must be adopted to segregate the data from various users.
- Jurisdictional and location policies must be established to govern the data location [15].

5.2   Protect Integrated Medical Devices

Solution Space and Its Challenges
This section covers the suitable mitigations, processes, and protection mechanisms. The aspects covered include reporting and feedback loops, risk management, regulation, and resilience activities which were considered important among others.

Complexity of the problem is increased when we consider the device functionality propagation. Security threats are evolving in nature and many unknown challenges exist [16].

5.2.1   Reporting and Feedback Loops

Medical device manufacturers and healthcare providers require good notification and feedback systems between them for addressing cyber security issues. Legislation to mandate reporting of cyber security incidents would assist in identifying issues from all healthcare providers. Auditing must become part of standard operational practice. It must be reportable to the organization's governance level [16].

5.2.2   Risk Management

Procedures, processes, and robust governance imply that the identification of risks and comprehending the management factors and incidental response are necessary. Regulatory compliance is additionally essential for the safety of the patient. Governance processes and risk management must include documenting data flows with respect to networked medical devices, ensuring that appropriate protection and safety is provided at every stage of data transfer, processing, and storage. This type of management can only be implemented using organizational policy, which is bolstered by effective procedures [16].

5.2.3   Regulation

When changes are made to the medical device including the embedded software, it must be subject to Food and Drug Administration (FDA) renewal, a process that costs time and money. This exposes known vulnerabilities longer than would occur otherwise and leads to extra expense for the manufacturer in the regulatory compliance process. The embedded code has inherent security vulnerabilities but is not of interest to the regulatory bodies. They are only concerned with the device. It was identified by the FDA Safety and Innovation Act (FDASIA) report that the FDA needed to be clearer in its aspects of regulation that will apply to cyber security vulnerabilities considering the increase in exchange of data between electronic medical record (EMR) systems and devices and the utilization of wireless spectrum [16].

5.2.4   Resilience Activities and Contingency Planning

Network segregation includes firewalls, limiting access, setting up virtual local area networks, and the use of uninterruptible power supplies on critical care devices.

All of these measures are essential to resilience and a standard part of contingency planning but have not completely considered medical devices as a part of the information system network. Contingency planning constitutes business impact analysis, incident detection and response, disaster recovery, and business continuity, in that order [16].

In case of a governance approach, there are three levels of organizational structure, all of which play a role in ensuring the safety of resources including medical devices and associated network. The strategic level includes policy development, compliance with regulation, and business process. The tactical perspective covers contingency planning, proactive approaches to risk management; education, and auditing. The day-to-day operational level involves daily practices such as implementing technical controls such as encryption routinely [16].

5.3 Enable the Secure Use of Shared Machines and Workstations

It must be ensured that multiple users can access machines and workstations across the organization. The users must be kept productive and given access to the data and tools anytime they need it.

Another important factor to consider is the physical security of shared machines and workstations. The system must be guarded from an unauthorized user as they may view sensitive data, destroy machinery/data, or steal machinery or data. Access and theft can be prevented by three main ways. The first two have been practiced for generations and comprise of using a human guard or a physical lock to protect the systems. The third method is one of importance because it limits access of a particular data set to a particular group of people/person based on biometrics, the answer to a security question, a password, a smartcard which bears a specialized chip, or a combination of more than one of the aforementioned.

These security measures coupled with the authentication and network controls mentioned previously in this chapter are together effective in maintaining security [17].

5.4 Control Identity Management and Databases

The type of authentication solution employed by an organization should be compatible with an organization's developing or existing IT infrastructure and at par with the amount of risk posed to a particular information system. The particular type of environment a person works in greatly impacts the mechanism of authentication used. A single mechanism cannot be applied to every field, even within the same jurisdiction. The feasibility of every authentication solution must be assessed with respect to public safety requirements and with the consideration that the authentication term needs to be parallel to the evolution of authentication technologies [18].

Local User Authentication

Local user authentication involves the input of a Personal Identification Number (PIN) or utilization of biometric reader by the user (e.g., fingerprint sensor, iris-scanning camera, microphone that authenticates speaker's voice) to obtain access to their mobile device, gaining access past a *"lockscreen"* [18].

PINs, Passwords, and Gestures

Passwords, gestures, and PINs are occasionally referred to as *"memorized secret tokens."* These tokens represent the current standard for local authentication, although this is gradually changing due the influence of biometrics. The dissatisfaction of users with using memorized secret tokens (e.g., passwords, PINs) on their mobile devices stems from the frequent entry errors and requirement of manually managing multiple PINs/passwords for a multitude of sites as well as portals. When public safety is considered, operational necessities may either constrain or prohibit the ability of a first responder to use a PIN, password, or gesture for authentication.

Physical Tokens

Physical tokens are a relatively uncommon type of local authentication, and they constitute something you have. These tokens generally take advantage of proximity and radio frequencies. Such proximity tokens are used to unlock a mobile device only when the token is within very close range to the device, similar to unlocking a car with the press of a button. The technologies used in such tokens include Bluetooth, Near-Field Communication (NFC), radio-frequency identification (RFID), etc., and could be worn on sleeves, as rings or elsewhere on the body of the user. Usually, the policies of the organization dictate how long the device remains unlocked as well as the required frequency of communication with the proximity token of the user.

Biometrics

Biometric tokens constitute something you are and are gaining more and more prominence as a form of local authentication. Biological/physiological characteristics that are unique to individuals can be used for authentication, such as the face, iris, palm prints, fingerprints, and voice but most of these characteristics are not often used along with mobile devices. Behavioral characteristics including aspects like the input texting pattern of a user can also be used for authentication. Accelerometers, gyroscopes, and other sensors included can determine gait, that is, walking rhythm/pattern and other behavioral characteristics. False Rejection Rate (FRR) and False Acceptance Rate (FAR) are measurements used to check and ascertain the correctness of the biometric system. This is usually done by introducing spoofs to the biometric system such as a picture of a person, slightly different fingerprint, etc.

5.5   Remote User Authentication

Passwords, smartcards, and biometrics are technologies that are used for remote user authentication for mobile devices. Remote authentication is distinguished from local authentication by the existence of many untrustworthy entities between the entity performing verification and the user. Sending information over an untrusted network is more commonly seen in remote authentication protocols [18].

PINs, Passwords, and Gestures
Passwords, gestures, and PINs for remote authentication are similar to those discussed under local authentication. These tokens are capable of attaining assurance level 1 or 2 at most and are often used in conjunction with cryptographic keys or biometric data to reach higher assurance levels.

Biometrics
The biometric authentication technologies available for remote authentication are similar to those available for local authentication for most of the part. The main difference occurs while using multifactor tokens with biometric information for local authentication. In this case, the verification process happens without any information leaving the token. However, while using remote authentication, verification occurs on backend systems that reside externally with respect to the mobile device. These backend systems provide greater computational power and thus, result in greater accuracy which in turn results in stronger authentication.

One-Time Password Devices
One-time password (OTP) devices are physical devices which are used to generate a password having a relatively short life span. These are generally used in the absence of an additional authentication factor and are employed in a similar manner to password entry. The OTP devices are more effective when coupled with any other form of authentication, such as memorized token, to provide multifactor protection.

Attached Contact Smartcard Reader
Smartcards consist of a processor capable of performing complex cryptographic operations and are generally used to store credentials. These devices are referred to as multifactor cryptographic tokens when used in conjunction with a PIN and are capable of reaching assurance level 4. Mobile devices generally cannot have smartcard readers built into them. This is due to the large size of the readers. Instead, an external smartcard reader can be used to access stored credentials. Smartcard readers are capable of being connected to mobile devices via Bluetooth, USB (Universal Serial Bus), or other interfaces for the reading of the stored credentials.

NFC Smartcard
The NFC smartcard readers are used to address the concerns regarding the usability of external smartcard readers. When the smartcard is placed within a few centimeters of an NFC-enabled device, the mobile device can communicate wirelessly with a smartcard and access its stored credentials. The card must be held in close proximity to the mobile device as the PIN is entered by the user. This

method provides multifactor authentication without the bulky external card reader mentioned before. Organizations that rely on NFC-capable devices must carefully choose their mobile devices to make sure of NFC-compatibility.

Software Cryptographic Tokens
Software cryptographic tokens are generally used in the absence of equipment to incorporate physical tokens and smart cards. These tokens are given protection by a memorized secret token and then stored within a mobile device's nonremovable internal storage. This increases the risk that the credentials could be stolen and therefore, hardware-based storage is preferred to software for the storage of credentials. Authentication is accomplished via the mobile operating system or in some cases, other external applications. Interfaces for storing and using software-based digital certificates are provided by all major mobile platforms.

Removable Hardware Security Modules
Removable hardware security modules are physical devices that provide trusted storage, encryption/decryption, and other cryptographic operations such as digital signatures. MicroSD and USB security tokens are common examples of this type and often consist of a processor that provides capabilities that are similar to that of a smartcard. These tokens can be used for the storage of sensitive information such as software cryptographic credentials while simultaneously providing resistance to tampering. Another example, universal integrated circuit card (UICC), resides within a mobile device and can be removed from the device with some extra effort. USB security and MicroSD tokens are comparatively much easier to remove and insert from a mobile device as long as it has the appropriate physical interface.

Embedded Hardware Security Modules
Embedded hardware security modules are similar to the removable hardware security type. The difference is that they cannot be removed from a mobile device. These types of embedded security modules are becoming more common in mobile devices. They are often in the form of distinct chips that are built into the device. The main advantage of these modules is that they provide authentication capabilities without external hardware. They also typically have the ability to store cryptographic keys securely and perform various cryptographic operations in hardware similar to their removable counterparts.

5.6 Remote Device Authentication

Remote device authentication is used in a manner similar to remote user authentication. Once provisioned, these devices can prove their identity to a verifier by proving knowledge of a particular credential. Establishment and management of a public key infrastructure (PKI) may be required by this approach and thus, the existing Federal PKI could be leveraged. If credentials were stored in storage locations that are hardware protected, a greater level of assurance would be achieved. The major difference from other forms of authentication is the lack of user interaction in providing a password or PIN for unlocking a credential for use [18].

5.7 AI-Based Solutions

One of the main solutions to AI security threat can be AI itself, as it can be used to detect cyber attacks. Extensive research has been done in this field. The success rate of those researches has varied between 85% and 99%. Some products have also been improved to detect cyber attacks with the help of Artificial Intelligence such as Darktrace.

5.8 Using Biometric Logins

Multiple cyber security experts believe that user personal information, social security numbers, and credit card information can be easily compromised via cyber attack on their password. Therefore, utilizing AI for cyber security has introduced the concept of biometric login techniques for secure logins. These systems can scan retina, palm prints, and fingerprints accurately and can be used in combination with passwords, PINs, etc., that are already used with devices like smartphones [19].

5.9 Detecting Threats and Malicious Activities

Conventional cyber security regimes make use of Advanced Threat Prevention for detecting cyber threats and providing protection against them. But, 845.37 million malware were created in the year 2018 and approximately ten million novel malware are created on a monthly basis. Traditional cyber security systems can handle known existing threats but are inefficient in handling such new varieties of malware. Therefore, one feasible solution is the adoption of AI [19].

AI systems are being trained by cyber security firms for the detection of malware and viruses with the help of several data sets that include codes and algorithms. Using such data, AI can perform pattern recognition that helps identify malicious behavior in software. Furthermore, machine learning can detect whether a particular website navigates to malicious domains by analyzing path traversals of websites and likewise can recognize malicious files and isolate them from the system preemptively. AI systems can be used in predictive analytics and in recognizing and comprehending the characteristics of a malware and then formulating appropriate response [19].

5.10   Learning with Natural Language Processing

Data can be automatically collected by AI-powered systems for reference by scanning studies, news, and articles on cyber threats. Natural Language Processing (NLP) can be used by AI systems for selection of useful information from the scanned data which provides information about cyber attacks and associated anomalies, mitigation, and prevention strategies. The end result of this analyzed information is the identification of timescales, calculation of risks, harvesting of data, and making of appropriate predictions [19].

5.11   Securing Conditional Access

Authentication models are used by organizations to secure important data from unwanted people/intruders. The system can be compromised via the network, if someone with higher authentication privileges is accessing the data remotely. Using AI in cyber security will help in creating a real-time, global, and dynamic authentication framework that modifies access privileges based on location/network. Multifactor Authentication can be used by AI systems for this purpose. The system collects user information to analyze the behavior of the application, user, device, data, network, and location using this approach. AI-powered systems can be used to automatically change any user's access privileges ensuring data security on remote networks, using such information [19].

6   Case Study

The case study that is discussed in this section has been developed by the authors based on credible news sources. The purpose of this case study is to describe an example of a recent cyber security mishap to give the reader a real-world understanding of how such a security threat is manifested, perpetrated by the criminal, and eventually remedied. We have take the example of a well-known company called TaskRabbit and broken down the various aspects of cyber security incident they faced in 2018.

TaskRabbit is a US-based mobile and online marketplace and was founded in 2008. It is believed to have greater than a million users across 40 cities in America, as well as London in the UK. It matches those who provide freelance labor services with local demand for such services. The business was acquired by Ikea in September 2017. Jesper Brodin (President and CEO of Ikea) wanted to tap into the digital expertise possessed by TaskRabbit while simultaneously providing customers affordable and flexible service solutions. The two companies engaged in a partnership that resulted in a new rapid service that provides help in furniture assembly to Ikea customers [20].

What happened?
TaskRabbit was taken down on the 16th of April, 2018 due to a cyber security incident. A few weeks after the incidence, the CEO of TaskRabbit, Stacy Brown-Philpot, revealed that the team had been *"working around the clock over the past few days and investigating"* the incidence. It was reported that preliminary evidence suggested *"Unauthorized user gained access"* to the systems of the company resulting in the compromise of certain identifiable information. The users were urged by the company to change their passwords immediately and monitor and report any suspicious activity. It was also advised to change the login credentials on other domains if the same password was used. The exact nature of the information accessed was not exactly known. The company was working alongside a forensics team to determine the individuals whose data were involved and the extent of breach. Such individuals were notified of the same [21] [22].

In the early occurrences of the security breach, users reported experiencing different kinds of technical glitches. One of these redirected users to WordPress upon attempting to access TaskRabbit. Another user indicated that she landed on the site through a phishing link that was received by email. Beyond that, much information was not provided regarding the nature of the breach, the means by which the attacker accessed the systems, the information stolen, or who the perpetrator was [22].

How Is TaskRabbit Dealing with the Situation?
The website was eventually back online. The company issued a statement trying to regain the trust of their clients and attempted to display openness and transparency. Additional security measures were taken upon its return including the evaluation of data retention practices for reducing the amount of data that the company stores, resetting of all the user passwords, and enhancement of network cyber threat detection technology, to prevent the recurrence of such events in the future. The CEO also reliably informed the public that measures to continuously update the security program were being taken under the premise that cyber threats evolve constantly. The company also had to collaborate with a team of law enforcement and cyber security experts to determine the specifications of the incident [20] [21, 22]. A depiction of the public address given by the company is provided in Fig. 5.

April 16, 2018

Dear TaskRabbit Community,

TaskRabbit is currently investigating a cybersecurity incident. We understand how important your personal information is and are working with an outside cybersecurity firm and law enforcement to determine the specifics. The app and the website are offline while our team works on this.

We will be back in contact with you with more information once we have it. As an immediate precaution, if you used the same password on other sites or apps as you did for TaskRabbit, we recommend you change those now.

Thank you for your patience while we investigate the issue and for being such an important part of our community.

-   TaskRabbit Team

Fig. 5 TaskRabbit's address to its community [22]

7 Future Prospects

With constant innovation and advancements in technology, the future of Artificial Intelligence seems to be quite promising. There are a large number of projects currently underway which will revolutionize the medical industry and the way in which research is being conducted today. These applications are solving the existing problems and ensuring better implementation. However, they are giving rise to other problems. What is thus clear, is that Artificial Intelligence and its applications will continue to advance but each stage comes with its own set of problems. As scientists, it is important we find solutions and stay ahead of the constantly evolving technology.

Some of the researches that are currently underway are as follows:

Neurochip This is a chip that can possibly allow for the communication to take place between a human tissue and an electronic device. This could thus steer the research toward integrated man machines and provide opportunities for integrating the living human brain and machines to control artificial limbs, monitor people's vital signs, correct memory loss, and possibly find cure for diseases like tumors and mental disabilities [23].

Health Monitoring
Better implementation of complex systems like Internet of things, big data, and machine learning has helped to improve the accuracy of data processing in health monitoring and e-health records.

Medical Robots
With the help of Artificial Intelligence, scientists are now developing medical robots that can be trained to perform very complicated and tedious jobs. They can be targeted to problematic areas that are inaccessible to a human. The most recent advancement is that of the computer-based surgery system that utilizes a medical robot to perform surgery. The doctor can thus sit at remote locations and perform surgery on a patient. At present, the medical robots are not employed for visible use cases in terms of fully automated surgery or other medical procedures. This is due to regulations that require that a recognized professional administer these procedures. Issues such as liability are harder to resolve with AI, as it is usually unclear how the AI-based system reached its conclusion [24].

8   Conclusions

Artificial Intelligence currently can be applied to varying degrees in various sections of healthcare including personalized medicine, drug development, patient diagnosis, treatment, and monitoring. Currently, one particular booming area is the analysis of radiological images for diagnoses/research using image processing in conjunction with AI. AI is also used for optimization processes on the administration side of healthcare. As mentioned in the previous section, AI has great scope for development in the future. The growth of AI technologies and their prevalence in healthcare are however greatly deterred by the inability to provide optimum levels of privacy and security to the patients and medical providers alike. This is an essential step in gaining public acceptance and trust in the technology. The utilization of AI on a large scale, having overcome its security concerns, can prove to be beneficial by saving costs, satisfying patients, and satisfying the requirement for staff/personnel. Thus, deliberate measures must be taken toward mitigating this risk. Preexisting solutions must be reviewed and improved upon; and novel solutions must be proposed. The purpose of this chapter was to delineate the various types of existing threats and their potential solutions in order to provide the reader with the basic information upon which further research can be conducted.

References

1. What Is Deep Learning? | How It Works, Techniques & Applications - MATLAB & Simulink. (2019). *What is deep learning? | How It Works, Techniques & Applications - MATLAB & Simulink,* [ONLINE]. Available at: https://in.mathworks.com/discovery/deep-learning.html#whyitmatters. Accessed 25 July 2019.
2. Schmidhuber, J. (2015). Deep learning in neural networks: An overview. *Neural Networks, 61,* 85–117.
3. Jackson, C. P. (1985). *Introduction to artificial intelligence: Second, enlarged edition (Dover 669 books on mathematics)* (pp. 21–24). New York: Dover publications. ISBN 0-486-24864-X.
4. Hinton, G., Deng, L., Yu, D., Dahl, G., Mohamed, A., Jaitly, N., Senior, A., Vanhoucke, V., Nguyen, P., Sainath, T., & Kingsbury, B. H. (2012). Deep neural networks for acoustic modeling in speech recognition – The shared views of four research groups. *IEEE Signal Processing Magazine, 29,* 82–97.
5. Clinical decision support systems | British Columbia Medical Journal. (2019). Clinical decision support systems | *British Columbia Medical Journal,* [ONLINE]. Available at: https://www.bcmj.org/articles/clinical-decision-support-systems. Accessed 25 July 2019.
6. The Medical Futurist. (2019). *The future of radiology and artificial intelligence – The medical futurist,* [ONLINE]. Available at: https://medicalfuturist.com/the-future-of-radiology-and-ai. Accessed 25 July 2019.
7. Shortliffe, E. H., & Cimino, J. J. (2014). *Biomedical informatics: Computer applications in health care and biomedicine* (4th ed., pp. 149–151). London: Springer-Verlag. https://doi.org/10.1007/978-1-4471-4474-8. ISBN:978–1–4471-4474-8.
8. Bashshur, R. L., Mandil, S. H., & Shannon, G. W. (2004). Telemedicine journal and e-health. *Telemedicine Journal and e-Health,* [Online], *8,* 3–4. Available at: https://doi.org/10.1089/15305620252933356. Accessed 27 May 2019.
9. Langlotz, C. P., Allen, B., Erickson, B. J., Kalpathy-Cramer, J., Bigelow, K., Cook, T. S., Flanders, A. E., Lungren, M. P., Mendelson, D. S., Rudie, J. D., Wang, G., & Kandarpa, K.. (2019). A roadmap for foundational research on artificial intelligence in medical imaging: From the 2018 NIH/RSNA/ACR/The academy workshop. *Radiological Society,* [Online], *291*(3), 1–11. Available at: https://doi.org/10.1148/radiol.2019190613. Accessed 28 May 2019.
10. PresCouter. (2019). *Artificial Intelligence meets drug development.* [ONLINE]. Available at: https://www.prescouter.com/2017/05/artificial-intelligence-drug-development/. Accessed 27 May 2019.
11. Tabatabaei, S. G. H., Dastjerdi, A. V., Wan Kadir, W. M. N., Ibrahim, S., & Sarafian, E. (2010). Security conscious AI-planning-based composition of semantic web services. *International Journal of Web Information Systems, 6*(3), 203–229.
12. Osoba, O. A., & William Welser, I. V. (2017). *The risks of artificial intelligence to security and the future of work.* Santa Monica, CA: RAND Corporation.
13. Awwalu, J., Garba, A. G., Ghazvini, A., & Rose Atuah, J. (2015). Artificial intelligence in personalized medicine application of AI algorithms in solving personalized medicine problems. *International Journal of Computer Theory and Engineering,* [Online], *7,* 1–5. Accessed 28 May 2019.
14. Sobia Hamid, D. (2016). The opportunities and risks of artificial intelligence in medicine and healthcare. *Communications Summer 2016 The Babraham Institute, University of Cambridge,* [Online], *1,* 1–4. Available at: http://www.cuspe.org/wp-content/uploads/2016/09/Hamid_2016.pdf. Accessed 28 May 2019.
15. Kumar, P. R., Raj, P. H., & Jelciana, P. (2018). Exploring data security issues and solutions in cloud computing. *Procedia Computer Science, 125,* 691–697.

16. Williams, P. A. H., & Woodward, A. J. (2015). Cybersecurity vulnerabilities in medical devices: A complex environment and multifaceted problem. *Medical Devices: Evidence and Research (Auckl), 8*, 305–316.
17. Zhang, Q., Qi, Y., Hou, D., Zhao, J., & Han, H. (2007). Uncertain privacy decision about access personal information in pervasive computing environments. In J. Lei, J. Yu, & S. Zhou (Eds.), *Proceedings of the 4th international conference on fuzzy systems and knowledge discovery (FSKD 2007)* (Vol. 3, pp. 156–160). Haikou, China: IEEE, 24–27 Aug 2007.
18. Hastings, N. E., & Franklin, J. M. (2015). *Considerations for identity management in public safety mobile networks*. NIST Interagency/Internal Report (NISTIR) – 8014.
19. Li, J.-h. (2018). Cyber security meets artificial intelligence: A survey. *Frontiers of Information Technology & Electronic Engineering, 19*(12), 1462–1474.
20. Chang, L., & Mogg, T. (2018). *TaskRabbit app goes down as it investigates 'Cybersecurity Incident'*. digitaltrends.com
21. Wyciślik-Wilson, M. (2018). *TaskRabbit returns after security breach and reveals 'personally identifiable information' was exposed*. betanews.com
22. Detrick, H. (2018). *TaskRabbit may have been hacked. What to know about the security breach that forced the platform offline*. fortune.com
23. *Brain-Machine, Artificial Intelligence and the Future of Biomedical Engineering Solutions for Health*. Retrieved on Sep 2018 from https://events.vtools.ieee.org/m/192547.
24. *Artificial Intelligence in Medical Robotics – Current Applications and Possibilities*. Retrieved on May 2019 from https://emerj.com/ai-sector-overviews/artificial-intelligence-medical-robotics/.

Design of a Low-Cost Sensor-Based IOT System for Smart Irrigation

Kunal Singh and Raman Kumar

1 Introduction

Water is the most precious resource in agriculture. As agriculture is fundamentally the most important sector of Indian economy, an appropriate measure to control and regulate constant supply of clean water at different intervals of time in a year is our utmost priority. It has been observed [1] through studies that the impact of climate change on the availability of water throughout the year is appreciable and has received much attention from scientific community in recent years. With ever-increasing population of India, the crop requirement would undoubtedly increase every year to feed this growing population, while the resources would remain limited. With the advent of new technologies and emerging sciences combined with recent researches, it is now possible to estimate the optimal resources required for a particular crop production, whether it is moisture, nutrients, or temperature of the field. It may be noted that in this world of competition, emerging technologies in the field of communication, artificial Intelligence, robotics, and actuation have flourished [2] beautifully and proved to be beneficial for the people of India and other developing countries. It is now possible to buy powerful computing devices inexpensively in these countries, and high cost is no longer a factor preventing the implementation of these technologies in smart irrigation systems. It may be noted that some of the Indian scientific communities, under ignorance and lack of

K. Singh (✉)
Mechanical Engineering Department, Maharaja Agrasen University-Baddi, Baddi, Himachal Pradesh, India

R. Kumar
Department of Computer Science and Engineering, I. K. Gujral Punjab Technical University, Kapurthala, Punjab, India
e-mail: er.ramankumar@aol.in; dr.ramankumar@ptu.ac.in

© Springer Nature Switzerland AG 2021
R. Kumar, S. Paiva (eds.), *Applications in Ubiquitous Computing*, EAI/Springer Innovations in Communication and Computing,
https://doi.org/10.1007/978-3-030-35280-6_4

adaptation to these emerging technologies, still perceive an advancement of technology with high cost and disregard the effective application of these smart irrigation systems in India. Such communities take for granted the availability of precious resource like water and neglect its scarcity in the near future. It may be noted that not only is the scarcity of water a threat for our crops, but also in a country like India where weather conditions turn around violently and storms and heavy rainfalls are prominent, an unregulated supply of water throughout the year has only proved to be a disaster for high-yield agriculture. It may be observed [3] that according to the United Nations (UN) Food and Agricultural Organization (FAO), our mother nature needs to produce around 70% more food by 2050, compared with its production in 2006 to not just satisfy but to barely feed our growing population. To achieve these goals, farmers as well as agricultural companies are diverting their attention to the Internet of Things [4] for analyzing the effective use of resources and achieving high production capabilities, so precision farming or smart agriculture is our only choice to flourish our future generation with adequate food. This chapter discusses a prototype of a smart irrigation system elegantly designed, considering the low purchasing capability of Indian farmers. A rope and pulley system has been designed to cover the crops with inexpensive canvases in case of storms or heavy rainfall. This canvas shed is controlled by an IOT-based control system, the design of which has been explained in detail in this chapter. The canvas shed can respond to different environmental conditions based on the analysis of the control system which takes decision depending on the data sent by different sensors situated in the field. This combination of low-cost mechanical actuation and sensory feedback control system in general can provide a sound platform for Indian farmers to begin their journey in precision farming and adapting IOT as a means for conserving precious natural resources, which may not be adequately available in the near future. The following sections explain the design and construction of the complete system in detail.

2   Literature Review

It may be noted that a framework [5] has already been set which utilizes the Internet of Things for smart irrigation. Such systems use heterogeneous devices, which are an integral part of the smart irrigation system and help monitor the system in real time. Also, the actuation of the mechanism is automated, based on the outputs of the sensors. There are also such systems [6] which aim toward autonomous monitoring of the irrigation for both large- and small-scale plantations using IOT. Such systems monitor the temperature and moisture content of the soil along with the detection of pollutants like $PM_{2.5}$ (particulate matter 2.5), PM_{10} (particulate matter 10), carbon monoxide, and NO_x (nitric oxide) in the air. Such an analysis helps in the estimation of a proper amount of fertilizer required by the crop and prevents unwanted toxicity to the soil. In another system [7], a process of autonomous irrigation in the urban gardens has been developed. It consists of multiple sensors and actuators in conjunction with Zigbee (A Wi-Fi shield for

Arduino microcontroller) for monitoring and controlling the system through a web portal in real time. The system can be automated, such that it prevents the watering of garden if there is a forecast of precipitation and the humidity of soil has not fallen below a critical value. In another system [8], a simple smart irrigation system has been proposed, which utilizes the solar energy for its operation and irrigation is automated using microcontroller. The status of irrigation is notified to the user's mobile using a Global System for Mobile Communications (GSM). It may be noted that such a system is simple and efficient in operation and does not require IOT technology. Another system [9], which utilizes a combination of GSM and IOT technology, proposes a design in which the temperature, moisture, and water level in the irrigation tank will be measured and sent to the user via GSM. The water pump is automated through messages on user mobile and an android application. The data of sensors are also being stored in cloud for analysis. In another system [10], a low-power consumption and long-distance transmission smart irrigation system has been proposed, which makes use of 6LOWPAN (Internet Protocol version 6 over Low-Power Wireless Personal Area Networks) to implement very low power transmission and networking. Another system [11] proposes a smart irrigation project, which is not only capable of monitoring the irrigation of a field but also checks the amount of the three major macronutrients, such as nitrogen (N), phosphorus (P), and potassium (K), of the soil. An ARM 7 Processor is used to control the system and all the sensors are connected to this controller. In this system, the user will be notified via an e-mail about the status of temperature, humidity, and macronutrient content of the soil.

In another system [12], a simple system is designed, consisting of Arduino UNO microcontroller board and ThingSpeak (open-source cloud platform) to sense the moisture content of the garden and to update regularly on the cloud in the form of a graphical representation. Such a simple system is effective and helps prevent the degradation of plants and helps maintain soil fertility for longer periods of time. Another author [13] proposes different models to supervise the moisture of soil to help Indian farmers in the irrigation of fields. Such models make use of several sensors and Raspberry Pi (a mini computer) to build intelligent systems for irrigation. In another system [14], open-source technologies are utilized to sense the soil parameters such as soil temperature and moisture to automate the irrigation of fields. This system also utilizes the data of weather forecast obtained through internet. The sensors involved also sense the amount of ultraviolet (UV) light radiation on the field. The system is based on the smart algorithm that utilizes the sensor data along with the data of weather forecast to decide and predict the requirement and level of irrigation in the field.

3 Design of a Smart Irrigation System

An IOT-based smart irrigation system must not only be capable of taking logical decisions based on sensory data, but must also be able to transfer these data reliably and wirelessly through cloud servers to any device connected to internet. The beauty of cloud computing lies in the fact that the data can be stored virtually on these servers and can be assessed by the user at any time or in real time. Figure 1 depicts the overall view of an IOT-based smart irrigation system. The system consists of three sensors, which are temperature sensor, moisture sensor, and rain sensor, out of which the temperature and moisture sensors gather information on the condition of soil while the rain sensor collects the information on rain.

The sensory data are sent to the IOT-based control system that wirelessly updates these data as well as the logical states of actuators to a mobile application connected to internet. It can be observed through Fig. 1 that the board is capable of taking smart decisions to protect the crop and can actuate the shed actuator and the water pump. The function of a shed actuator is to cover the crop with a shed in case of heavy rainfall or to cover the crop to prevent any unwanted evaporation from soil in case of dry and hot environmental conditions. It may be noted that all the logical conditions, which actuate the shed, have been fully illustrated in Fig. 2 representing the flowchart of the complete system. In Fig. 2, the middle portion of the flowchart represents the logical conditions under which the shed will actuate. As there are three sensors determining the condition of soil, the decision taken by the control system will depend on various combinations of these sensors' output. Therefore, the middle section of the flowchart can be divided into three parts. According to the program, one of the decisions is that the shed will be actuated when the moisture of the soil exceeds the preset defined value and it is heavily raining, in this case, the shed will get closed. It may be noted that this condition is independent of the temperature data. The idea behind this condition is to prevent the drowning of crops under heavy rainfall. The other condition is based on the temperature sensor data alone, that is when the temperature of the soil exceeds a defined value then the

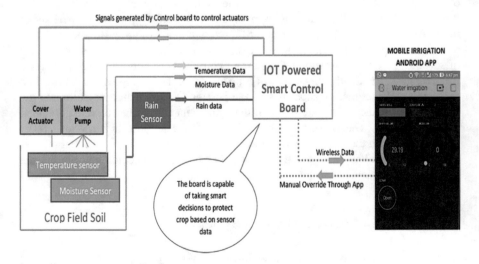

Fig. 1 Block diagram of IOT-powered smart irrigation system

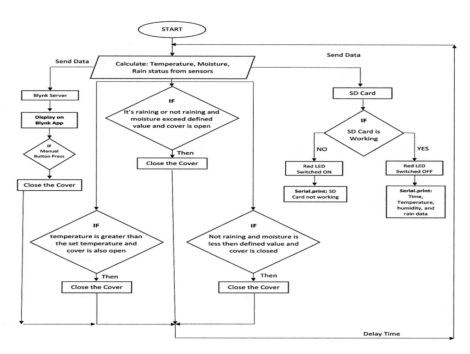

Fig. 2 Flowchart of IOT-powered smart irrigation system

shed will be actuated and get closed. This decision prevents the evaporation of soil under excessive heat of the sun. The other condition prevents the evaporation of moisture under minimum rain condition. In this case, the shed will be closed when it is not raining and the moisture is below the set value. It may be noted that the three decisions defined above represent the conditions which can be programmed in the control board based on the requirements of the farmer, depending on the environmental factors required by the crop. These conditions represent an intelligent attempt at the creation of an unmonitored smart irrigation system and can be expanded or deleted based on the personal preference of the farmer. As no change in hardware of the system is required to change these conditions, the system itself can be customized at very low cost. This feature makes this smart control board adaptive to a variety of different regions in India and also compatible with the crops grown in different seasons requiring different environmental conditions. It can be observed from Fig. 2 that the sensory data as well as the status of field actuators can be monitored on mobile application connected to internet. This is the core IOT feature of the smart control board, as the system can not only be monitored from any part of the world but the crop shed can be closed or opened by the farmer using the manual override button in the mobile application based on the visual indications of the sensory data. The application also helps the farmer to keep a check on the functionality of this smart irrigation system.

From Fig. 2, we can observe that sensory data are being stored in a memory card integrated in the control board. This feature allows the researchers to analyze the effect of environmental conditions on the production of crops; therefore, this system cannot only help improve crop production but also determine the factors that need to be controlled for better crop production.

4 Hardware Arrangement of a Smart Irrigation Control System

Figure 3 represents the block diagram of the smart control board. The board receives the data from the sensors and stores them simultaneously on a Secure Digital (SD) card. The control board also controls the outputs of the relay board and motor driver board, which controls the water pump and shed situated in the field. It may be observed that the relay board and motor driver boards are not integrated within the control board itself. The main reason to keep them separate is to enhance the adaptive capabilities of the control board. The power required by the water pump and field shed actuators depends on the wattage of the motor used, which, in turn, is different for different farmers. A small farmer growing medicinal plants might have preinstalled or preferred a small wattage motor for pumping water which will definitely save his money, while a farmer with large crop fields requires a high-wattage motor for irrigation. Therefore, the relay board can be replaced easily with high- or low-wattage relay board or even solid-state relays or Mosfets (metal–oxide–semiconductor field-effect transistors), etc., without affecting the working or design of the control board. Figure 4 represents the actual circuits of this smart irrigation control board in operation.

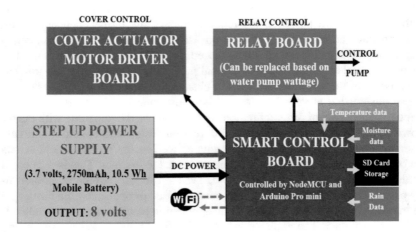

Fig. 3 Block diagram of smart irrigation control board

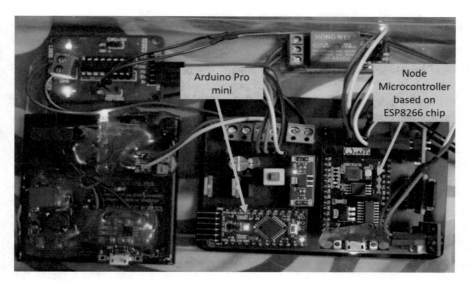

Fig. 4 Actual smart irrigation control board

The memory unit of the control board is a combination of two microcontroller boards of different architecture working in harmony with each other. The following two control boards are used:

• NodeMCU
• Arduino Pro Mini

The following sections describe the operation and working of these two micro-controller boards in this smart irrigation control board.

4.1 NodeMCU Microcontroller Board

The main control unit of this smart irrigation control board, which is equipped with inbuilt IOT capabilities, is NodeMCU [15]. It is an open-source IOT platform that contains a firmware running on an ESP8266 Wi-Fi system on a chip made by Espressif Systems [16]. The hardware of NodeMCU is based on the ESP-12 module. Originally, the microcontroller unit (MCU) needed to be programmed in Lua script. It may be noted that Arduino, which is an open-source software and hardware community which is popular for designing control boards based on the Atmel series of microcontrollers, has recently made changes in its IDE (Integrated Development Environment), so that its IDE can program NodeMCU. It may be observed that as DIY (Do It Yourself) community as well as researchers, hobbyists and electronic enthusiasts are more familiar with Arduino IDE for the development of embedded systems than the Lua script. With the capability of Arduino IDE to program ESP8266 controllers in its native Arduino C language, new doors were opened for the DIY community to explore this controller to its depth. The NodeMCU, used in this smart irrigation control board, has been programmed in Arduino C, which makes the programming much more adaptable and easier to understand for debugging purpose. NodeMCU and all the ESP8266-based microcontrollers work on 3.3-volt power supply; therefore, the sensors, such as temperature, moisture, and rain sensors connected to MCU, work on this 3.3-volt power supply.

4.2 Arduino Pro Mini Microcontroller Board

It may be noted that it is possible to connect a sensor pin to different microcontrollers and it does not affect the quality of data output from the sensor. This little trick expands the multitasking capabilities of an embedded system, that is, different operations can be performed based on the sensor data at the same time without indulging in interrupts. Considering the above advantages, the sensory pin of the moisture sensor, apart from being connected to NodeMCU, is also connected to Arduino Pro mini microcontroller board. A program fed in the Arduino Pro mini microcontroller board operates the water pump motor in real time by turning it on or off based on the moisture content of the soil, while the same moisture sensor simultaneously sends the data to NodeMCU which transmits it via Wi-Fi to an IOT application for display purposes. Also, as stated previously, both NodeMCU and Arduino Pro mini must work at the same voltage level, which is 3.3 volts in this case for the moisture sensor to deliver the exact readings to both these microcontrollers. Therefore, a 3.3-volt input supply is required for Arduino Pro mini. It is not advisable to use this 3.3-volt supply from NodeMCU itself as Arduino Pro mini is also connected to a relay board and any voltage surge, even for a millisecond, can disrupt the operation of NodeMCU. Therefore, a separate 3.3-volt power module (ASM1117) is used to supply this power, the input to this module is 5 volts and is supplied by a separate 7805 voltage regulator (5-volt step-down regulator) dedicated for this purpose only.

5 Detailed Overview of Sensors and Other Components

It may be noted that the most important components of the designed control board are the sensors, from which all the data of the soil are being transferred to the control board itself. The board consists of the following three sensors that transmit the sensory data to the NodeMCU and Arduino Pro mini.

- 3.3–5 volt raindrop (waterdrop) detection sensor
- 3.3–5 volt soil moisture detection sensor
- DS18B20 digital temperature sensor

5.1 Raindrop Sensor

It is a resistive sensor and is basically an easy way to detect the rain. It works on the principle of change in resistance between two metal wires when in contact with a common conductive liquid medium. This sensor consists of two parts, one is the base plate with a mesh of closely placed copper traces with little gap between them. The other part of this sensor is the operational amplifier board which has two outputs: one analog and other digital. As the resistance of the base plate changes when in contact with conductive medium such as rainwater, the output resistance from this plate is fed to the operation amplifier board which gives an output high or output low from the digital pin of the board. The board also consists of an LED (light-emitting diode) to give visual indication of the sensor output. When rain falls on the base plate, the resistance between the copper traces decreases and the output of the digital pin of the control board goes low which was initially high during no rain. The analog output of the amplifier board gives a relative measure of the amount of rain on the base plate. The sensitivity of the digital output can be adjusted using an on-board potentiometer. Figure 5 depicts the base plate and control board of the rain sensor. While the base plate is situated in the field, the board is connected to the base plate via wires which in turn is situated inside the smart irrigation control board, as shown in Figs. 3 and 4. The working voltage of this sensor is 3.3–5 volts depending on the control board type. If this sensor is attached to Arduino, then the base voltage provided by Arduino is 5 volts, but in this case, it is being connected to NodeMCU, whose base voltage is 3.3 volts.

5.2 Soil Moisture Detection Sensor

The control board of this sensor is exactly similar to that of the raindrop sensor, except the base plate whose shape is a bit different. The working principle of this

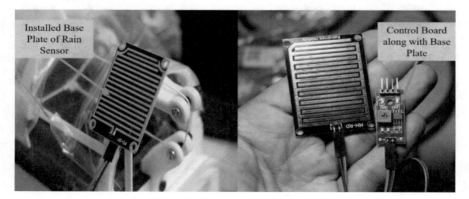

Fig. 5 Base plate and control board of rain sensor. Note that the control board is an electronic device and is needed to be kept separate from rain or water

sensor is also based on the change in resistance. It may be noted that the working current of both rain sensor and soil moisture sensor is less than 20 mA, so they consume less amount of power. This sensor comes in two flavors: one resistive and other capacitive. The accuracy of capacitive sensor is more as compared to resistive sensor, but this system is constructed using a resistive sensor, as it is quite inexpensive and serves the purpose of this system well. The resistive soil moisture sensor consists of two probes, which can be used to estimate the content of water in terms of its volume. The current is passed from one probe to another using soil as its medium. When the soil is wet, the resistance decreases and it indicates more water content in soil. It may be noted that the dry soil is a poor conductor of electricity therefore the resistance will be more. The main drawback of the resistive soil moisture sensor is the corrosion of probes. This drawback is solved using capacitive sensor which is made of a corrosion-resistant material. Figure 6 depicts the soil moisture detection sensor used in this system.

5.3 DS18B20 Digital Temperature Sensor

This is a digital temperature sensor [17] which is the waterproof version of the original sensor. It has an inbuilt 12-bit analog to digital converter. It can be easily connected to an Arduino digital input. It communicates with the microcontroller

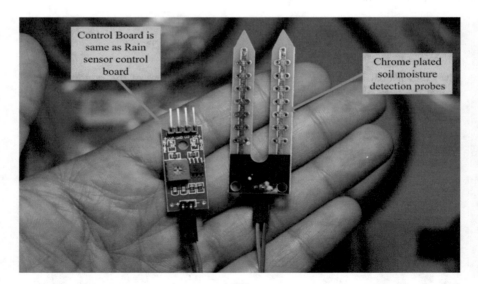

Fig. 6 Soil moisture detection sensor; note that the control board is the same as the rain sensor control board

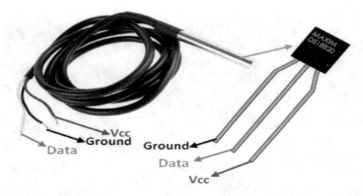

Fig. 7 Waterproof digital temperature sensor used for temperature sensing of soil. (Image downloaded from https://components101.com/sensors/ds18b20-temperature-sensor on May 29, 2019)

using a single wire. It consists of three wires: ground wire (GND), input voltage wire (VCC), and data wire. It can work within a voltage range of 3.3–5 volts and has a temperature measuring range of $-50\ ^{\circ}C$ to $+125\ ^{\circ}C$. The accuracy of this sensor is $0.5\ ^{\circ}C$. This temperature sensor has a wide range of applications and is usually used to measure the temperature in adverse environments. The waterproof version of this sensor is robust in nature and is appropriate to measure soil temperature in the field. Figure 7 depicts the temperature sensor used in this system.

It may be noted that the relay board, to drive water pump, is kept separate from the control board to make the system adaptable to the requirement of the farmer. The specifications of the relay board, shed actuator motor, and its driver as well as the mini water pump used are as follows:

5.4 Five-Volt Relay Board

This type of low-end relay board is mostly used to build prototypes and test the circuit. The relay module used here can switch AC (alternating current) as well as DC (direct current) power. Output AC rating of this board is 250 volt AC, 10A and the DC rating is 30 volt DC, 10A. It may be noted that such power rating is sufficient to run most of the lightweight pumps, but is not appropriate for heavy-duty pumps. The input to this module is simply connected to the data pin of the microcontroller. Figure 8 depicts the relay module used in this system.

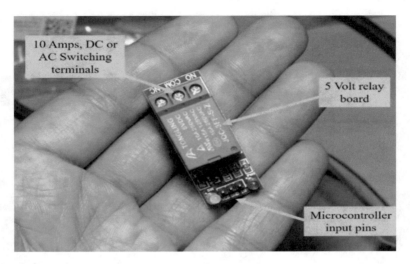

Fig. 8 The term 5-volt relay implies that it can be triggered directly via microcontroller pins working at the base voltage of 5 volts

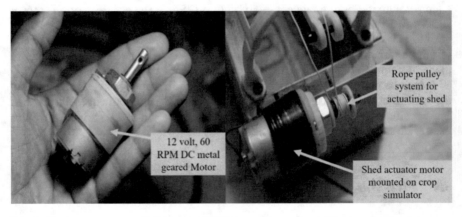

Fig. 9 Shed actuator motor (DC 12 volt, 60 RPM-geared motor) used for the crop simulator

5.5 Twelve-Volt 60 RPM Direct Current Motor

This type of DC motor is inexpensive and is used for testing and prototyping a control system. Figure 9 depicts the motor which is used to actuate the crop simulator. It is a geared motor with nylon or metal gears. The motor used in this system consists of metal gears.

Fig. 10 L293 motor driver board. (Image downloaded from https://images-na.ssl-images-amazon.com/images/I/51wtGN3-mfL._SX522_.jpg, courtesy of Amazon.com)

5.6 L293D Motor Driver Board

This module is used to run medium power motors and is suitable for driving stepper motors or DC motors, as used here in this system. The integrated circuit (IC) used in this board is L293D which is an H-bridge driver. The board requires four data lines from the microcontroller which, in this case, is NodeMCU. It also requires a ground and 5-volt power connection from the microcontroller. The power to run the motor is separate from the power to drive the IC from microcontroller. This board can supply at most 12 volts with a DC current limit of 0.6A, which is sufficient to run the actuator motor used in this system. Figure 10 depicts the motor driver board used in this system. It may be noted that this H-bridge driver board can drive two DC motors simultaneously and has the capability to change the direction of motors.

5.7 Mini Submersible Water Pump

The mini submersible water pump is mostly used in small projects to simulate irrigation. Here, it is used to test the control of the smart irrigation control board. It may be noted that the voltage rating of this pump is 3–6 volt DC and the working current is around 130–220 mA. It consumes little power, ranging from 0.4 to 1.5 watts and is suitable for such simulation and testing of this smart irrigation system. The flow rate of this pump is around 80–120 liters per hour. Figure 11 displays this motor used for the simulation of irrigation.

As discussed, all the sensor data are being stored on an SD card. An SD card reader module holds the SD card that stores the data. The specifications of the module are as follows:

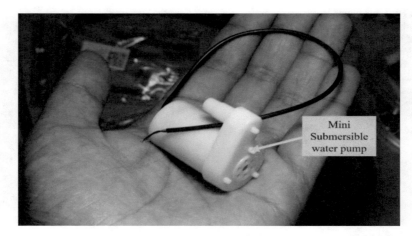

Fig. 11 Mini submersible water pump used for pumping water in the plants of the simulator to maintain the moisture level

5.8 SD Card Module

This module is interfaced with microcontroller via serial peripheral interface [18]. It consists of data pins like MOSI, MISO, SCK, CS and power pins like VCC and GND. The data from the sensors are being transferred through synchronous serial communication, which is suitable for short-distance communication, especially in embedded systems. This module works on a 3.3-volt power supply, which is being supplied directly from NodeMCU. Figure 12 depicts the SD card module used for this system. The control board is programmed, such that the sensor data are stored in the excel file in the form of a table. The control board is programmed, such that the time interval between sensor data readings, which are being stored in the SD card, can be stored in a text file within the SD card itself. The program stored in the control board reads the time interval from this text file itself.

Another important part of the system is the power supply for the control board. Complete specifications of this control board are discussed below:

5.9 Power Supply

The complete system, that is, the control board, sensors, relay board, water pump, shed actuator, and motor driver, is being powered by step-up power supply developed from a mobile lithium-ion battery whose specifications are 3.7 volts, 2750 mAh, which is around 10.5 Wh. The voltage is stepped up using a step-up

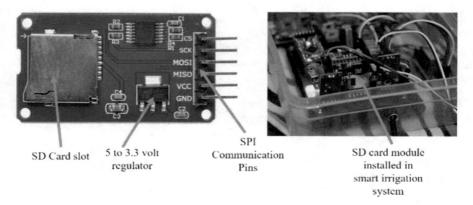

SD Card slot   5 to 3.3 volt   SPI   SD card module
                regulator   Communication   installed in
                                    Pins   smart irrigation
                                                 system

Fig. 12 SD card module used in smart irrigation systems

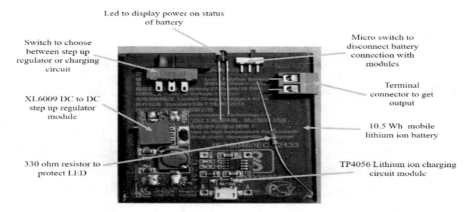

Fig. 13 Components of power supply system

regulator (XL6009 DC to DC step-up module), also a lithium-ion charging circuit based on IC TP4056, which is being used to charge the battery when discharged. This power supply system is specifically developed for this smart irrigation system, to serve the power requirement of the system for long hours. It may be noted that in actual conditions the power supply for water pump and shed actuator and its driver has to be different due to high-wattage requirement. Figure 13 depicts the disassembled view of the power supply system developed for this smart irrigation system.

This system is designed and developed in EagleCAD software [19]. It is a well-known reputed printed circuit board (PCB) designing software from AutoDesk, which is an American multinational software corporation. Figure 14 displays the schematic of this system in detail. Figure 15 depicts the single-sided PCB board layout of the smart irrigation system.

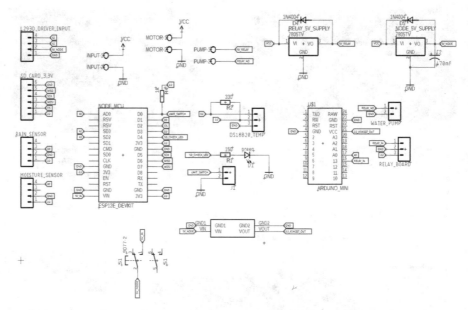

Fig. 14 Actual schematic diagram of IOT-based smart irrigation control board developed in EagleCAD software

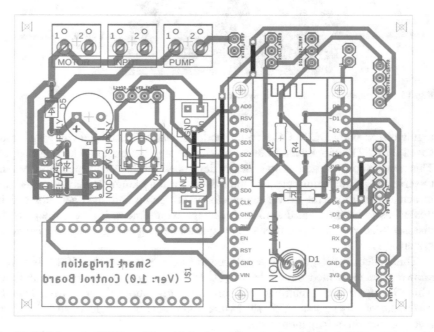

Fig. 15 PCB layout of IOT-based Smart irrigation system developed in EagleCAD software

6 Mechanical Components of the Crop Simulator

Figure 16 depicts the computer-aided design (CAD) model of the crop simulator. The idea is to develop a mechanical system that is inexpensive and effective. It should also be a system that can cover the crop rapidly and is reliable. A miniature prototype of the actual mechanical system has been developed to mimic the system, which can be applied in the field itself. It may be observed that the rope and pulley mechanism, which can drive a canvas rapidly, that is, open or close it, is quite effective as well as inexpensive. In the CAD model, railings or smooth rods as shown serve the purpose of guide rails for the canvas. An angle slide helps move the canvas on the guide rails and motor mounted on the shed actuator motor mount helps move the angle slide bidirectionally. As one end of the canvas is fixed to the mounting and the other is fixed to the angle slide, a bidirectional movement of the angle slide causes the canvas to open and close rapidly with the movement of angle. It may be noted that this mechanism is not only inexpensive but also easy to install as well in medical fields or other plantation areas where high-value crops or plantation is grown and it is very important to protect the field from natural disasters.

The parts of this miniature crop simulator, such as rope pulley brackets, motor mount, and mountings, etc., are three-dimensional (3D) printed from acrylonitrile butadiene styrene (ABS)-based filament using the fused deposition modeling technology. The infill [20] (amount of plastic within the part) of these parts is kept to 100% for maximum strength, as these parts have to bear the tension of ropes during the motion of angle slide and canvas shed. It may be noted that in reality these parts will be constructed from wood or metal for large fields for optimum strength and reliability. However, the future is not far ahead when such large and complex

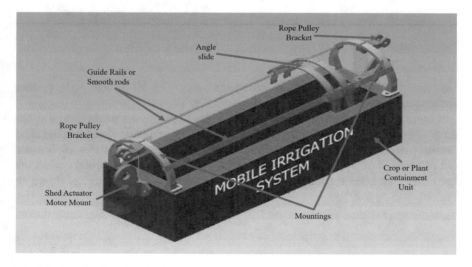

Fig. 16 Crop simulator mechanism developed in Autodesk Inventor Professional software

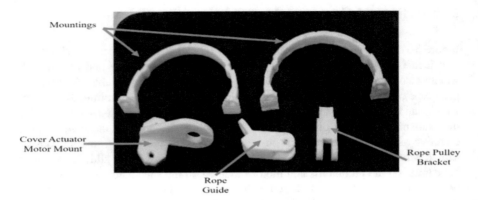

Mountings

Cover Actuator
Motor Mount

Rope Pulley
Bracket

Rope
Guide

Fig. 17 3D printed crop simulator parts printed using 100% infill

Crop Shed   Guide Rails

IoT based Real Time
Smart ___ gation System

Plants in the
Crop simulator

Real time data
being received on
smart phone
through IOT

Fig. 18 Actual simulator with plants

structures can be printed using 3D print technology. The 3D printed parts, printed for the miniature crop simulator, are shown in Fig. 17.

Figure 18 depicts the actual simulator developed to test the working of a smart irrigation control system. It may be noted that for this miniature simulator, the shed is composed of a thin plastic sheet and an angle slide, as discussed in Fig. 16, is made out of cuttings of a polyvinyl chloride (PVC) pipe. The environmental conditions are being controlled for the plants grown in this miniature simulator, as shown in Fig. 18.

7 Cost Estimation and Implementation of Design

The cost associated with this system can be broadly classified into two parts. One being the cost of control board and the other being the cost of mechanical system required to implement the design. It may be noted that the cost of control board and sensors is fixed and may be minutely increased, based on the requirement of few additional sensors depending on the field area. However, the cost of a mechanical system employed in the field can vary drastically depending on the field size. For medicinal or experimental plantation, the mechanical system cost is quite low. However, for large fields a robust and low-maintenance mechanical system to control the shed needs to be employed. Such systems may require a high-torque Synchronous or AC induction motors as an actuator to run the shed mechanism. Also for large systems, relays are required, instead of motor drivers. It may be noted that simple smart irrigation systems are already available in the Indian market as well as globally. However, the technology of irrigation involving IOT has not yet reached its market potential in India and is still a topic of research and analysis. The total cost of the smart control system, as proposed in this chapter, is around 1500/− rupees which can be further scaled down under mass production. Comparing the cost with that of a similar system such as "SMARTDRIP" [21], available in the Indian market with an MRP of 4399/− rupees, it may be observed that the proposed system is cost-effective.

8 Pros and Cons of the System

Every innovative system when first designed for practical implementation in the market faces some difficulties and objections. Such systems surely have some advantages as well as disadvantages. The system as proposed is IOT based and is designed to make use of cloud servers to send and receive control signals to control the system. Such a control system requires Internet connection at both ends for operation. It may be noted that in a country like India, the Internet connection may not be available always in the fields, especially during heavy rainfall or storms. However, it is expected that in a growing economy like India, this problem will be totally eliminated in the near future. The advantage of having IOT as a means of communication is that the system can be operated from anywhere around the globe. The disadvantage, however, is the nonavailability of Internet connection, as discussed above. Looking at the cost, the system has an advantage of being inexpensive, considering the technology involved. It may be noted that the current system, as proposed, is not IP66 certified and needs to be encased in IP66 standard encasing for weather- and waterproofing. Such casings are easily available in the market and are inexpensive. Also, the circuits and components involved need to be of Industrial grade for additional reliability and long life. Such implementations need to be considered before actual mass production of the system.

9 Conclusion

In this chapter, we discussed the design and construction of a smart irrigation system which is IOT based and inexpensive in nature. In the upcoming future, where resources will not be plentiful, such systems can act as a good choice for an Indian farmer to understand and farm in a scientific way. As the technology is becoming inexpensive day by day and India is the leading customer of mobile phones, it is not only good in terms of effective use of technology but is also inexpensive for an Indian farmer to save the destruction of crops using this advanced technology. It may be concluded that with the advancements in virtual and sensor technology, it is fairly easy to build such smart irrigation devices and even easier to use these devices. It may be concluded that the IOT can play an important role in the design and development of such systems. With the help of IOT devices, it is possible for a farmer to not only monitor the status of his crops but also control and prevent the crops from unwanted destruction from anywhere in the world. It may be observed that unleashing the true potential of IOT in the field of agriculture is not only one of the best ways to save precious resources like water in the upcoming future but also to regulate the proper supply of these resources in accordance with the requirement of crops. With more advancements in such a smart irrigation system, it may also help farmers to grow crops in adverse conditions. Such a development may require the use of artificial intelligence in combination with IOT and is the subject of future research.

References

1. Islam, A., & Sikka, A. K. (2010). Climate change and water resources in India: Impact assessment and adaptation strategies. In M. K. Jha (Ed.), *Natural and anthropogenic disasters*. Dordrecht: Springer.
2. Sivagami, A., Hareeshvare, U., Maheshwar, S., et al. (2018). Automated irrigation system for greenhouse monitoring. *Journal of The Institution of Engineers (India): Series A, 99*, 183. https://doi.org/10.1007/s40030-018-0264-0.
3. World must sustainably produce 70 percent more food by mid-century -UN report. (2018, December 8). Retrieved from https://news.un.org/en/story/2013/12/456912
4. Why IoT, big data & smart farming are the future of agriculture. (2018, December 8). Retrieved from https://www.businessinsider.com/internet-of-things-smart-agriculture-2016-10?IR=T
5. Koduru, S., Padala, V. G. D. P. R., & Padala, P. (2019). Smart irrigation system using cloud and internet of things. In C. Krishna, M. Dutta, & R. Kumar (Eds.), *Proceedings of 2nd international conference on communication, computing and networking* (Lecture notes in networks and systems) (Vol. 46). Singapore: Springer.
6. Dasgupta, A., Daruka, A., Pandey, A., Bose, A., Mukherjee, S., & Saha, S. (2019). Smart irrigation: IOT-based irrigation monitoring system. In M. Chakraborty, S. Chakrabarti, V. Balas, & J. Mandal (Eds.), *Proceedings of international ethical hacking conference 2018* (Advances in intelligent systems and computing) (Vol. 811). Singapore: Springer.

7. Caetano, F., Pitarma, R., & Reis, P. (2015). Advanced system for garden irrigation management. In A. Rocha, A. Correia, S. Costanzo, & L. Reis (Eds.), *New contributions in information systems and technologies* (Advances in intelligent systems and computing) (Vol. 353). Cham: Springer.
8. Bhuvaneswari, C., Vasanth, K., Shyni, S. M., & Saravanan, S. (2019). Smart solar energy based irrigation system with GSM. In L. Akoglu, E. Ferrara, M. Deivamani, R. Baeza-Yates, & P. Yogesh (Eds.), *Advances in data science. ICIIT 2018* (Communications in computer and information science) (Vol. 941). Singapore: Springer.
9. Meeradevi, S. M. A., Mundada, M. R., & Pooja, J. N. (2019). Design of a smart water-saving irrigation system for agriculture based on a wireless sensor network for better crop yield. In A. Kumar & S. Mozar (Eds.), *ICCCE 2018. ICCCE 2018* (Lecture notes in electrical engineering) (Vol. 500). Singapore: Springer.
10. Jiang, X., Yi, W., Chen, Y., & He, H. (2018). Energy efficient smart irrigation system based on 6LoWPAN. In X. Sun, Z. Pan, & E. Bertino (Eds.), *Cloud computing and security. ICCCS 2018* (Lecture notes in computer science) (Vol. 11067). Cham: Springer.
11. Raut, R., Varma, H., Mulla, C., & Pawar, V. R. (2018). Soil monitoring, fertigation, and irrigation system using IoT for agricultural application. In Y. C. Hu, S. Tiwari, K. Mishra, & M. Trivedi (Eds.), *Intelligent communication and computational technologies* (Lecture notes in networks and systems) (Vol. 19). Singapore: Springer.
12. Guchhait, P., Sehgal, P., & Aski, V. J. (2020). Sensoponics: IoT-enabled automated smart irrigation and soil composition monitoring system. In M. Tuba, S. Akashe, & A. Joshi (Eds.), *Information and communication technology for sustainable development* (Advances in intelligent systems and computing) (Vol. 933). Singapore: Springer.
13. Das, R. K., Panda, M., & Dash, S. S. (2019). Smart agriculture system in India using internet of things. In J. Nayak, A. Abraham, B. Krishna, G. Chandra Sekhar, & A. Das (Eds.), *Soft computing in data analytics* (Advances in intelligent systems and computing) (Vol. 758). Singapore: Springer.
14. Goap, A., Sharma, D., Shukla, A. K., & Rama, K. C. (2018, December). An IoT based smart irrigation management system using machine learning and open source technologies. In *Computers and electronics in agriculture* (Vol. 155, pp. 41–49). Elsevier: ScienceDirect.
15. Bajrami, X., & Murturi, I. (2018). An efficient approach to monitoring environmental conditions using a wireless sensor network and NodeMCU. *Elektrotechnik und Informationstechnik, 135*, 294. https://doi.org/10.1007/s00502-018-0612-9.
16. Official NodeMCU Datasheet. http://espressif.com/sites/default/files/documentation/0a-esp8266ex_datasheet_en.pdf
17. Qi, S., & Li, Y. (2012). The design of grain temperature-moisture monitoring system based on wireless sensor network. In M. Zhao & J. Sha (Eds.), *Communications and information processing* (Communications in computer and information science) (Vol. 289). Berlin, Heidelberg: Springer.
18. McRoberts, M. (2010). Reading and writing to an SD card. In *Beginning Arduino*. Berkeley, CA: Apress.
19. Official Eagle CAD information. https://www.autodesk.com/products/eagle/overview
20. Samykano, M., Selvamani, S. K., Kadirgama, K., et al. (2019). Mechanical property of FDM printed ABS: Influence of printing parameters. *International Journal of Advanced Manufacturing Technology, 102*, 2779. https://doi.org/10.1007/s00170-019-03313-0.
21. SMARTDRIP Automatic WiFi drip irrigation water timer-works with SmartPhone, Google Home and Alexa. https://www.amazon.in/SMARTDRIP-Automatic-Irrigation-Timer-Works-Smartphone/dp/B07D68NMNG?tag=gooinhydr18418-21. Information of product obtained from the given link on 25th July, 2019 at 10.00 A.M.

An Overview of Clinical Decision Support System (CDSS) as a Computational Tool and Its Applications in Public Health

Praveen Kumar Gupta, Abijith Trichur Ramachandran, Anusha Mysore Keerthi, Preshita Sanjay Dave, Swathi Giridhar, Shweta Sudam Kallapur, and Achisha Saikia

1   Introduction

The world today has become synonymous with the presence of large amount of data due to the perpetual rise in the human population. Drastic changes in the environment have led to multiple epidemics, drug and antibiotic resistance, superbugs, etc. Demographically, distinct health concerns are steadily mounting as well. In order to tackle these difficulties, novel technologies that can aid the human brain in efficient diagnosis are the need of the hour. One such technology is the clinical decision support system, commonly abbreviated as CDSS. A specialized algorithm is built into the system that helps in the generation of patient-specific suggestions. The roots of CDSS can be traced back to the very beginning of medical informatics. The earliest known record of the origins of CDSS can be found in a 1959 paper based on the logic behind a physician's reasoning. The paper described a probabilistic model for medical diagnosis, based on the set theory and Bayesian inferences. The primitive systems were mainly based on broad user inputs, and therapy suggestions were based on them [1]. The paradigms of CDSS have constantly changed over the past few decades with newer research coming into the picture. Clinical decision support systems commonly work in tandem with an electronic health record (EHR). The EHR is a database that contains a patients' medical records. The EHR can be queried with a keyword to search and retrieve the

P. K. Gupta (✉) · A. M. Keerthi · P. S. Dave · S. Giridhar · S. S. Kallapur · A. Saikia
Department of Biotechnology, R.V. College of Engineering, Bangalore, India
e-mail: praveenkgupta@rvce.edu.in

A. T. Ramachandran
Department of Computer Science, R.V. College of Engineering, Bangalore, India

© Springer Nature Switzerland AG 2021
R. Kumar, S. Paiva (eds.), *Applications in Ubiquitous Computing*, EAI/Springer
Innovations in Communication and Computing,
https://doi.org/10.1007/978-3-030-35280-6_5

patient-specific data. The CDSS algorithm is then applied to these data to obtain recommendations on the course of treatment.

The intricate structure of the clinical decision support system ensures its effective implementation. Data mining in conjugation with relevant clinical research may be employed to examine patient medical records in a CDSS. The methodologies used in the implementation of CDSS are detailed below:

Fuzzy logic rule-based approach: Fuzzy logic mimics the human decisions as per the inputs given to the system. Since all the medical data need not lie in state of extremes such as absolute truth and absolute false, fuzzy logic allows the ability to compute data in the form of relative property and thus enhances the understanding, severity and decision logics in a much more specific manner, similar to the workings of a human brain.

The Bayesian networks: Any diagnosis, prognosis, treatment options, etc. are uncertain in nature. These problems can be tackled effectively using probabilistic methods such as Bayesian network. This type of CDSS is considered to be the most effective method for managing uncertainty.

Rule and evidence-based systems: Rule-based systems are fully based on human crafted rules or curated rules, which can also serve as the basis for decision-making and help in the mapping of the neural network.

Evidence-based systems are based on the computation of results based on the previously existing evidential data, and this helps in the estimations of use of drugs and treatments on a person, which is completely relative to the past results on decisions.

Genetic algorithms: They are roughly based on survival of the fittest, and the solutions to a processed outcome are further tested on various medical parameters to find the best suitable procedure for a given problem.

Artificial neural network: It is a modelling technique used in building relationships by extracting and analysing the pre-historic data, to understand the relationship between the input and the outcome of the processed data. Since no analysis and relationship can be extracted for the processed outcome, usage of such a method is highly unaccountable. Ergo, with further research, this could prove to be a viable method.

Hybrid systems: It is the interconnection of systems between the components responsible for physical inputs and properties with the systems responsible for analysis and computation for capturing live medical data, such as pulse rate and blood pressure, which is captured in analogue systems, and these act as the inputs for decision-making with the incorporation of fuzzy logic or artificial neural network to provide a solution or a suggestion for further processes to be required.

Three methods are primarily used to distinguish clinical decision support system for enhancing clinical practice functioning as a single variable analyses, direct experimental information, and multiple logistic regression analyses. Clinical outcomes of CDSS include morbidity studies, mortality studies, screening for cardiovascular disease (CVD) risk factors, antimicrobial treatment decisions, mental health, substance abuse treatment, and various other outcomes. Medication and prescription dosage levels can be cross-checked and verified in order entry systems, giving clinicians access to the right information at the right time and reducing errors which may occur due to the illegibility of medical prescriptions. CDSS combined with EHR has the capability to reduce rehospitalizations when used to aid pharmacogenetic testing for patients older than 50 years who often take multiple medications and experience adverse drug events (ADEs). It is not only beneficial to patient care, but it also results in potential health resource utilization savings. It has also paved the way for policymakers to improve the quality of health care and reduce costs. A meta-analysis of 148 randomized control trials to evaluate for clinical outcome improvement and cost reduction with clinical decision support revealed with strong evidence that CDSS usage can improve process outcomes like increased cautionary services with an odds ratio of 1.42 and increased ordering of apt medical treatment with an odds ratio of 1.57.

EHR can be defined as "an electronic version of a patient's medical history, which is maintained by the provider over time, and may include all the key administrative and clinical data relevant to that person's care under a particular provider, including demographics, progress notes, problems, medications, vital signs, past medical history, immunizations and laboratory data." [2]. Clinical decision support systems are often consolidated with EHRs to smoothen workflow and acquire the benefits of existing data sets. The incorporation of the EHR system, which uses real-time data to ensure high-quality patient care, with CDSS has the potential to revolutionize the health-care industry.

CDSS can be used by clinician's pre-diagnosis, diagnosis or post diagnosis in the form of a diagnosis decision support system (DDSS), which proposes appropriate diagnosis based on patient data, or a case-based reasoning (CBR) system, which utilizes previous case data to determine appropriate treatment options. CDSS can be knowledge based, where a pre-compiled, updatable knowledge base is combined with the patient's medical history to provide appropriate results, or it can be non-knowledge based where machine learning is employed to find patterns in clinical data and make suitable recommendations.

CDSS market share distribution globally with respect to three continents shows that North America accounts for 72% of the market share, with US holding 92% of shares, followed by Europe – 15% of global share, leading countries being France, Germany, and the United Kingdom. Asia accounts for 8% of shares, leading countries being Japan and China. The rest of the world accounts for only 5% of the total global share. There is, thus, an immense potential for the application of this system in underdeveloped and developing countries, where there is a higher prevalence of nutritional diseases and epidemics and decreased resources to tackle these issues. CDSS can be utilized to improve the standard of health care in such countries, having a major impact in terms of access to standard health care and effective treatment options. According to new market intelligence estimates, the value of the CDSS market in India is at USD 43.8 million. The sector is expected to grow to USD 206.1 million by 2025, by which time the total health-care IT market of India will reach a USD 2530.3 million valuation (BIS Research, Global Big Data in Healthcare Market-Analysis and Forecast, 2017–2025).

Skilful usage of the clinical decision support system can enhance the quality of health care, improve physician performance and amplify clinical performance for drug dosage. For example, it has proved to be a blessing in the field of palliative care where a clinical decision support tool called palliative care outcome scale (POS) has been proven to be a multidimensional measure to assess patient symptoms and conditions over long periods of time. The clinical decision support system has many forms, enabling it to accept multiple parameters, thereby making it a highly adaptable system. However, this system is not without flaws. CDSS is still at its infancy stage and has to be explored and developed further to extract the maximum use out of it. Most of the CDSS tools being released in the market are unidimensional. Multidimensional forms of CDSS are patient specific and provide rapid decision-making. They are based on real-time patient data such as medical history, physical examinations and laboratory data. Clinical prediction rules and evidence-based algorithms are able to accurately predict the patient's diagnosis, prognosis and the probability of the likely response to treatment. But the implementation of these tools is highly complex, making the rate of adoption of these tools very low. The drawbacks that exist due to various factors of the clinical decision support have been extensively delineated in this chapter.

CDSS, thus, serves as an efficient system to monitor and improve global health-care scenario, and if implemented effectively, clinical decision systems represent the future of health care.

2 Significance of CDSS

A CDSS aids and enhances clinical practice by effectively boosting the standard, nature and safety of the world health-care system. A CDSS software assists fundamental care providers and clinicians by presenting well-timed data to viably diagnose a patient by efficaciously pointing out to the actual health problem, reducing misdiagnosis, medication errors and elevating the quality of the care patients receive.

A handheld clinical decision guide system implemented into computers acts as an active knowledge system, which makes use of patient precise information to produce case-specific advice. This system keeps a record of the medical history of the patient and recommends a patient-specific treatment.

CDSS has reliable mechanisms and tools which can provide reminders, rec-ommendations and databases to store the data related to a particular patient for protection and preventive care. They can also supply constant alerts on probable dangerous drug interactions. The usage of CDSS can decrease expenses and increase efficiency and alert medical practitioners on ineffective testing. This practice enhances patient protection by avoiding potentially unsafe and high-priced complications, reducing patient inconvenience.

Furthermore, CDSS allows for more accurate medication information and dosing calculations that can be adapted into the clinical setting without much difficulty and significantly changing the pre-existing work structure enabling accurate evaluation of patient-specific statistics promptly resulting in an appreciably improved clinical practice.

Advantages of using CDSS over manual systems are several. For example, it is built into the medical process from the starting till its completion in a proper workflow rather than its integration in a discrete screen. The fact that CDSS is digital as an alternative to paper-based makes it more time saving and convenient to the practitioner. In addition to that, it offers real-time data, and hence, the information is offered at the current time and location as opposed to checking the patient earlier but providing the result later. Conclusively, a clinical decision support system presents recommendations for care rather than just assessments to be filled.

3   Benefits of CDSS

Patient Safety
The safety of the patients is improved by high involvement of system-initiated advice that provides safety information to the patient, for example, alerting the care providers to the seriousness of drug interactions, contraindications (a situation where drugs or any medical procedure could be harmful to a patient) with different drugs and awareness against prescribing medications for infants and elderly. Additionally, it also involves sending out messages to patients like time duration of therapy and forms of drugs.

Medical Errors
CDSS can be an integral tool in decreasing medical errors. This is done by helping out the practitioners in averting negative drug reactions and lowering unsuitable drug dosing to a patient. CDSS minimizes error rates by monitoring human errors by the medical practitioner's behaviour by offering numerous approaches to decision support, which includes caution, reminders, evaluation and guidelines for improving health care.

Basic or unsophisticated clinical decision support provides information and recommendations on drug doses, time interval and frequencies. Sophisticated clinical decision support performs functions such as drug allergic reaction checks, drug laboratory value checks, and drug–drug interaction checks. Two of the most noticeable functions are it provides reminders about drug guidelines such as advising the person about the necessary information while taking up the drug to avoid adverse effects and sequence orders, for example, causing the person to order glucose checks after he or she has ordered insulin [3].

Diagnostic and Workflow Process
CDSSs show great potential in minimizing clinical diagnostic errors and enhance the quality of medical care. Computerized CDSS assists practitioners to confirm

that they meet the necessities of long-term care. These computerized systems are programmed to examine a person's trait and condition and provide suggestions regarding the upcoming diagnosis.

It is imperative to plan and implement a beneficial CDSS so that it improves and enhances a physician's workflow. CDSS results in an ideal and reliable system that produces an acceptable and smooth system performance. Additionally, implementing CDSS depends on important factors such as finance, clinician ethics, leadership and management [4].

4 Structure and Applications of CDSS

4.1 Fuzzy Logic-Based CDSS

Fuzzy logic is derived from artificial intelligence and is used to represent the mathematical modelling of linguistic terms (variables) and human comprehension of knowledge. This enables the computers to solve problems and tackle problems similar to medical professionals.

Inaccurate data which can be vague and contain many uncertainties can be dealt by fuzzy logic which is typically difficult to be comprehended and understood by humans, which results in a high probability of error in the medical procedure if not observant.

The decision support system created using fuzzy logic is relatively successful in processing linguistic data compared to conventional sequential systems. Since the problems found around us majorly comprise uncertainties and vagueness, it proves to be extremely difficult to represent such problems into a mathematical model, which can then be used to build sequential algorithms to solve such problems [5].

Fuzzy logic helps in providing a deterministic conclusion based on multiple information, that is, the input of the experts and the biomedical sensors in order to arrive at a distinct solution.

In simple words, the structure of the fuzzy logic is nothing but IF C_1 AND C_2 THEN S, where C_1 and C_2 represent the condition cases for the event S or operation S to occur. These are used to express terms of linguistic data members (variables). Since the complexity of the system and architecture increases for mathematical formulas, therefore linguistic terms are preferred.

The number of rules and logic required for a complex computation generally depends on the number of inputs, outputs, and also on the goal of a computation based on the designers control response [6].

The usage of this type of approach proved to be useful in various medical operations/procedures such as injecting customized or tailored amount of anaesthetics required for patients, and this would result in a minimized dosage of the standard amount injected in patients. It also has the potential of decreasing the workload of

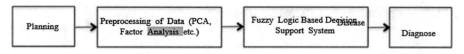

Fig. 1 Clinical decision support system with fuzzy logic [5]

the anaesthetics injected and thereby achieve the preferred depth of the anaesthesia used. It could provide clinical support to the medical professionals in maintaining a consistent and adequate number of anaesthetics to be used during a surgery, thereby giving time and room for the professionals to carry out high prioritized tasks instead of diverting attention on other tasks required to be performed simultaneously (Fig. 1).

Fuzzy logic is used due to its inaccurate reasoning; therefore, all logic dictates truths that are half done or relatively accurate in terms of the standards and structures defined for the fuzzy logic.

It dictates the human capabilities like human reasoning and judging uncertainty.

Since fuzzy logic is based on inaccurate reasoning, we refer human reasoning to be interpolative reasoning as the process neither lies in complete truth nor complete false but lies somewhere in between. This approach requires fuzzy logic to compute partial truth unlike in the conventional cases systems only processes complete Boolean output, that is, 1 s and 0 s.

4.1.1 Proposed Systems

Following the above ideology, a fuzzy-based clinical decision support system can be built on the illustrated design below. The proposed system consists of components such as rule-based system, inference engine, fuzzifier and defuzzifier. Different biomedical sensors feed their output to the inputs connected to the fuzzifier. The fuzzifier is responsible for the conversion of the outputs of the different sensors into quantifiable information that can be read and understood by the inference engine. The inference engine then activates and applies the rules on the data obtained. The observed and expert knowledge is saved in the rule-based system. The rule base system consists of all the fuzzy logic quantification, which gives the idea or the method to obtain good control of the current event. Linguistic fuzzy terms are nothing but the medical professional's knowledge on the occurrence of an event. The inference machine is responsible for correlating the data obtained from the fuzzifier and the rule base system in order for interpreting the description of the input based on the rule. The defuzzifier then converts the output of the inference engine into an appropriate human comprehendible language [6] (Fig. 2).

4.1.2 Applications in Clinical Studies

Pulmonology

Chronic obstructive pulmonary disease (COPD) is a disease caused due to exposure of mustard gas. It has many negative effects, such as cancer and pneumonia, and causes early as well as late complications. The chemical-injured victims majorly suffer from chronic respiratory complications. In order to provide extensive care and remedies for the patients, the fuzzy systems were used. One of the most common methods is the Mamdani fuzzy inference model, which controls a combined steam engine with a set of linguistic rules gathered from the patient's past experiences [7].

The measurement of lung disability is based on the Spiro metric measure, which is represented in percentage. The COPD is calculated with the aid (Fig. 3).

Fig. 2 Block diagram of the proposed system [6]

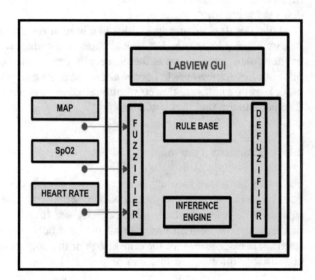

Severity (percentage)	Spirometry	Classification
0	FVC>80/FEV1>80 65<FVC<80	Passive lung disease
5-20	65<FEV1<80 50<FVC<65	Mild
25-45	50<FEV1<65 40<FVC<50	Moderate
50-70	40<FEV1<50	Severe

Fig. 3 Classification of pulmonary disease based on Spirometry [7]

NO	Variable		Rules	Linguistic Label	Fuzzy interval
1	inputs	FEV_1	$FEV_1>80$	VH	75–85–95–100
			$65<FEV_1<80$	H	60–70–80–85
			$50<FEV_1<65$	L	45–55–60–70
			$40<FEV_1<50$	VL	0–10–45–55
2		FVC	$FVC>80$	VH	75–85–95–100
			$65<FVC<80$	H	60–70–80–85
			$50<FVC<65$	L	45–55–60–70
			$40<FVC<50$	VL	0–10–45–55
3		FEV_1/FVC	$FEV_1/FVC>80$	VH	75–85–95–100
			$65<FEV_1/FVC<80$	H	60–70–80–85
			$50<FEV_1/FVC<65$	L	45–55–60–70
			$40<FEV_1/FVC<50$	VL	0–10–45–55
4	output	Severity	At risk	At risk	0–1–2–3
			Mild	Mild	2–3–4–5
			Moderate	Moderate	4–5–7–8
			Severe	Severe	7–8–10–11

Fig. 4 Overview of membership functions [7]

After the inputs are processed by the fuzzifier, these variables are used for determining the extremity in the decision-making table. The classifications of severity are as follows:

(i) Forced vital capacity (FVC)
(ii) Forced expiratory volume (FEV)
(iii) FEV_1/FVC ratio

Figure 4 gives the membership functions of the input and output fuzzy variables.

On the analysis of the membership functions, it was inferenced using the Mamdani fuzzy inference system (Figs. 5 and 6).

Cardiology

Fuzzy logic can be extensively used in the risk prediction of heart patients. The process consists mainly of two phases:

(i) Generation of the weighted fuzzy rules by an automated approach.

In order to obtain the weighted fuzzy rules, it requires attribute selection of attributes, attribute weighting mechanism and data mining.

(ii) Developing a fuzzy rule-based decision support system.

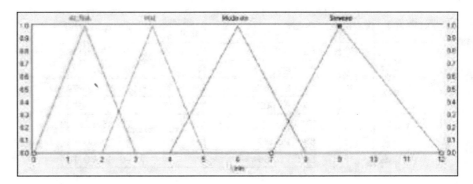

Fig. 5 Output of the membership functions for severity variable [7]

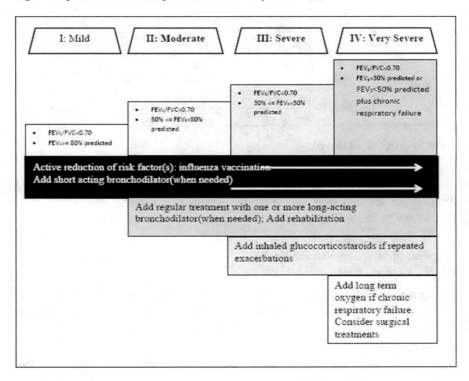

Fig. 6 Therapeutic recommendations [7]

The disease dataset of the heart may contain some noisy information and missing values; hence, there is a need for data pre-processing in order to remove the noise and missing values. The input database is divided into two subsets using class label, which are required for the mining of frequent attributes. On using the deviation range which are computed using the frequent attributes, the construction

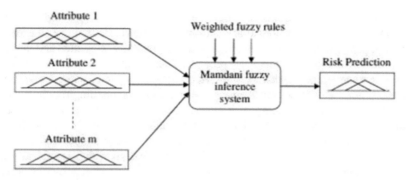

Fig. 7 Fuzzy inference system designed based on weighted fuzzy rules [8]

of the decision rules is made and these are scanned in the learning database to the corresponding frequency. The weighted fuzzy rules are obtained from the frequency, and on using the Mamdani fuzzy inference system, risk prediction fuzzy is developed, which proposes or produces the statistical data of the risk (Fig. 7).

4.2   The Bayesian Network

The conventional approaches to model casual or relationship between different diagnostics, which are not just a simple logical sequence of a rule, are implemented by a first-order logic. This is impractical due to the presence of incomplete knowledge and exhaustive rules that cannot be applied to solve the problems. The incomplete knowledge can be either theoretical due to the lack of advancement in science or practical due to the lack of data.

Bayesian networks can be used to model these types of casual relationships using degree of belief which are used to enable the system to compute reasonable under uncertainty. On using Bayesian networks, complex decisions can be represented using a directed acyclic graph.

The represented graph may contain all the relevant information required whose random variables may be connected by any abstract or casual dependencies responsible for decision-making [9].

Each variable represented in the graph contains a conditional probability table which indicates the probabilistic cause or influence of the variable. From the graph obtained, we can determine the patient's characteristics and other factors which can be termed as evidence. These evidences are computed by different inference algorithms which are then sent throughout the network on the consideration of the probabilistic occurrence for all the existing unobserved variables present within the same network. The unobserved variables can be any undiagnosed patient's condition, treatment performed or therapy alternatives. If required, we can further analyse it using different algorithms to identify any findings which are extremely influential or any variables that have not been observed yet having a high diagnostic value. Due to the fact that clinical distribution support systems require considerable amount of readjustments, they generally fail to be clinically integrated and thus are a source and termination for university projects. There is always a good probability of generating exaggerated results by a Bayesian network system, which could render the system unreliable if not given enough resources and modelling required [10].

For the efficient functioning of a clinical decision support system, expert and detailed modelling is required, which is costly, and for complex designs, the level of detail lies between the information which is extremely simple but useless to medical professionals and extremely detailed information which is typically hard to model and computed. However, the case may be the Bayesian networks describe a decision based on the joint probability distribution over a specific number of possible events. Depending on the number of variables having a possible direct influence, the number of required parameters used for indicating a single variable and its associated network can increase exponentially.

Due to the above reasons, the models need to be modelled with an extremely high grade of detail, which would allow derived graphs to be simplified to the most associable variables required. This type of processing and modelling is required as there exists restrictions in data for learning probabilities.

Bayesian networks are powerful tools that can be used to aid the clinical decision due to its probabilistic inference (Fig. 8):

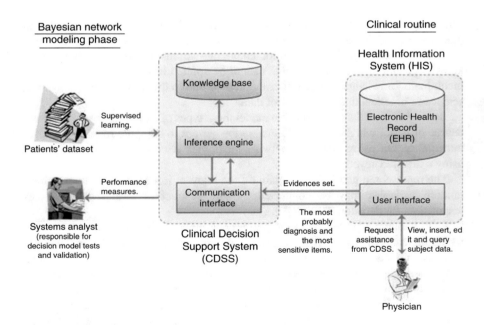

Fig. 8 Clinical decision support system components [11]

1. It can be easily understood by a clinician or a medical professional due to the inherent graphical nature of the networks output.
2. They can formally incorporate prior knowledge while learning the structure and parameters of the network.
3. Since the joint probability has a compact representation, they help in parameter estimation.
4. They allow observational inference and causal interventions.
5. They are more versatile when compared with classifiers that are build based on specific outcome variables and hence can be used to query any given node in the network.
6. They perform well in making predictions with incomplete data, since the predictor variables are used to estimate not only the query variable but also one another provided that the degree of modelling and associations are of high grade [12].

4.2.1 Applications in Clinical Studies

Diagnosis of Dementia, Alzheimer's Disease, and Mild Cognitive Impairment
Dementia generally involves symptoms affecting the memory and thinking abilities. It causes memory loss in patients, and Alzheimer's disease is nothing but the extreme progression of dementia. The memory loss could be due to variety of factors such as damage or loss of nerve cells and their associated connection.

On modelling a Bayesian network using a combination of data-oriented modelling and professional knowledge, it showed better results in the diagnosis of dementia, Alzheimer's disease and other mild cognitive impairments when compared to other known classifiers.

It also states the contribution of factors to the corresponding diagnosis, thus providing additional useful information to the clinical professionals [11].

Lung Cancer Care
Lung cancer is caused due to cigarette smoking. It is a malignant lung disease and often has no symptoms until the severity is complicated and advanced. General treatment for lung cancer is chemotherapy, surgery, radiation and targeted drug therapy. Therefore, it is very difficult to predict the survivability of a patient and hence causes extensive uncertainty.

The lung cancer experts can be aided by providing them with treatment selection suggestions and patient-specific survival estimates which can be done through Bayesian networks since they are effective in reasoning with the uncertainty domain [12].

Predicting Post-Stroke Outcomes
A stroke is a medical condition which results in poor amount of blood flow to the brain, hence causing death of brain cells. It is one of the most common causes of death and is also the leading cause of long-term disability.

Using Bayesian network, the mortality of different patients having stroke could be predicted with consequently a smaller number of risk variables in obtaining the risk rate, thus enabling medical professionals to provide better care and alternatives for the patients [13].

4.3   Belief Rule-Based Architecture System

Generally, an architecture of a system is nothing but how different components consisting of inputs, processes, and outputs are organized. The style of an architecture is the layout or pattern of system organization.

The proposed belief rule-based architectural system consists of the following layers:

- Interface layer:

  - This layer deals with the interaction between the user and the system. The interface layer is responsible for the data to be fed in by the user and proposed or suggested output to be displayed. In the below proposed system, the interface facilitates acquiring the leaf nodes that is the antecedent properties of the belief rule-based data. These data consist of clinical data, medical data, signs, symptoms, etc. By taking the account of belief of the domain expert, the data are distributed over the referenced values that are associated with the antecedent properties.

- Application processing layer:

  - The application processor deals in the interaction with the training module and the inference engine. The belief rule-based inference system consists of various number of components, for example, transformation of the inputs, rule activation weight calculation, updating of rule and the aggregation of the rules.
  - The evidential reasoning algorithm deals with the aggregation process of the inference engine.
  - The inference engine of this model works by first reading the input from the interface layer. The input data are then transformed into readable or referential values of the belief rule-based model antecedent properties. Then, all the activation weights of the belief rule-based model are calculated, which is then updated to the belief degree of the consequence. In rules and on using the evidential reasoning algorithm, all rules are aggregated.
  - The subsequent layer is solely responsible in building the training module by finding an optimal parameter and also by reducing the variation in the system results and the sampled data.

- Data management layer (Figs. 9 and 10):

  - The data management layer is based on the belief rule-based, clinical facts and other medical data. In order to develop the belief relief-based knowledge base, any of the following procedures can be used to implement.

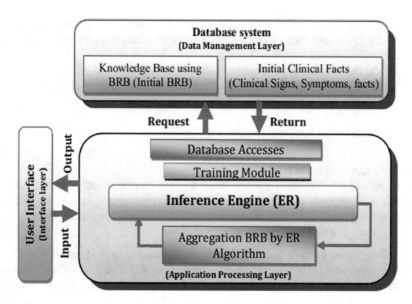

Fig. 9 Design level BRB CDSS architecture

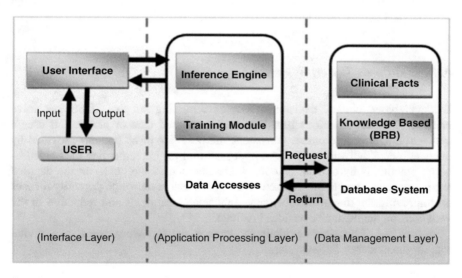

Fig. 10 Implementation level BRB CDSS

- Using the expert knowledge, the belief rules can be extracted from it.
- On the examination of previous or historical data, patterns can be generated and extracted for the belief rules.
- On the extraction of previously developed belief rules.
- Usage of arbitrary rules without any prior knowledge or data.

4.3.1 Logic-Based Systems in Clinical Studies

Assessing the Suspicion of Heart Failure
Any structural or functional cardiac disorder that causes inability of the heart to function normally can be termed as heart failure. This results in very short breath and, in many cases, can cause death. It occurs due to coronary artery diseases and usually affects the elderly in majority of the cases.

On using signs, symptoms and risk factors of patients, a belief rule-based clinical decision support can be constructed, which uses RIMER that allows handling of variety of uncertainties thus making it a robust and efficient tool. Usage of this system has not only caused reduction of costs in many lab investigations and assessments but also facilitate patients in taking the necessary precautionary steps. It has found to be more reliable and informative compared to the traditional cardiologists' suggestions [14].

4.4 Evidence-Based System

Evidence-based clinical decision systems are nothing but the CDSS that solve or tackle medical situations based on evidence-based data or practice. It uses a systematic approach to ensure consistent and best care of all patients provided by the health-care delivery system. An important tool for implementing the evidence-based practice is by clinical practice guidelines, which are based on the detailed expert consensus and latest research on the specific domain which primarily focuses on improving the diagnostic accuracy, enhancing treatments and reduction in the variations on the medical decisions involved.

In the case of physician guideline adherence, the following factors play a vital role for the evidence-based system:

- Awareness
- Familiarity
- Agreement
- Self-efficacy
- Outcome expectancy

Different field of medicine has its own set of factors or combined set of factors from adjacent domains.

On this concrete foundation, it has shown that there is a reduction in practice variability and substantially, providing a more satisfiable outcome for the patients. The clinicians recommend the CPG to be low, which is around 3–5 despite the benefits involved. Policymakers are encouraged by the use of IT to translate research findings into practice.

Therefore, when clinical decision support systems are applied to the evidence-based logic, they are referred as evidence adaptive. The goal of this system is to bridge the gap between practice and evidence and has shown great potential in achieving this [15].

Evidence-based system in nursing has proven to be a crucial aid for the best possible care, as they provide evidence-based recommendations to the nurses during the procedure or an event.

Clinical practice guidelines can be implemented, which has proved to be effective (Fig. 11):

- Remainder system
- Academic detailing
- Combined interventions
- Interventions that deliver accurate and patient-specific advice on real time and space

4.4.1 Applications in Clinical Studies

Nursing
Nurses have viewed the clinical decision support system to be of critical use as it improved interdisciplinary communications and helped their decision-making and self-confidence on treating the patients during any emergency crisis. The system has also enabled them to access information on best practice, which therefore enables them to provide a consistent care [16].

4.5 Artificial Neural Networks

Overview of artificial neural networks
Artificial neural network (ANN) is a popular type of non-knowledge-based CDSS. Non-knowledge-based CDSS uses machine learning rather than user-based knowledge. Neural networks learn by example, they cannot be programmed to perform a particular task. Neural networks are inspired from neurons that are present in our bodies. An ANN works the same way the biological one does. ANNs simulate human thinking by evaluating and eventually learning from existing examples/occurrences. ANNs consist of artificial neurons which are known as perceptrons.

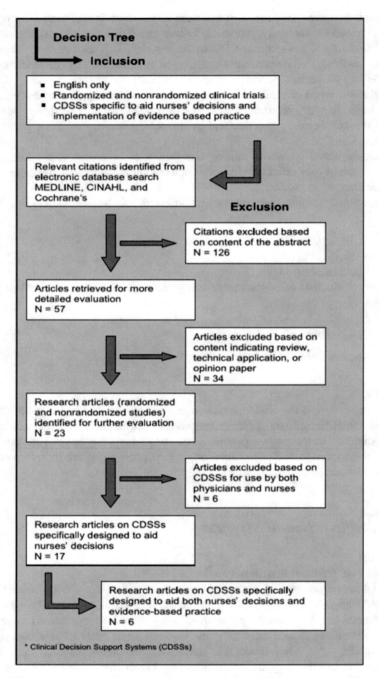

Fig. 11 Inclusion and exclusion criteria for literature search on evidence adaptive CDSSs in nursing

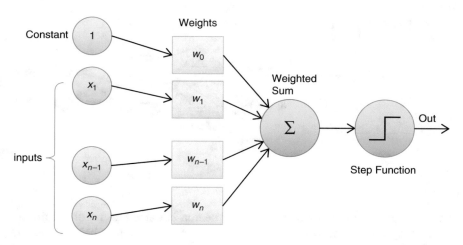

Fig. 12 Diagram depicting a perceptron

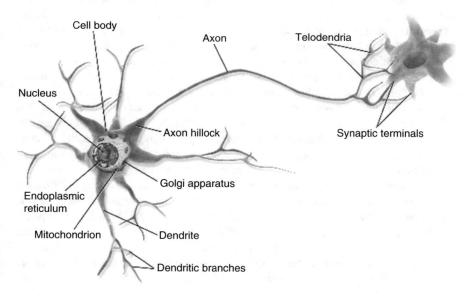

Fig. 13 A biological neuron

Figures 12 and 13 explain the working of ANN. Multiple inputs $x_1, x_2 \ldots x_n$ are fed to the network. Each input has its own corresponding weights, that is, w_1 for x_1, w_2 for x_2 and so on. Next, the weighted sum of these elements is calculated and then passed through a step function which is a type of activation function. The step function provides a threshold value above which the neuron will fire. Finally, the perceptron contains outputs. There are two modes in a perceptron: training mode

Fig. 14 Model of a neural
network

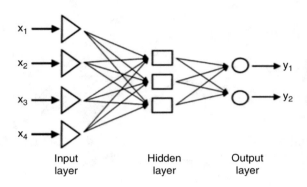

and using mode. In the training mode, the neuron is trained to fire or not fire for distinct input patterns; this essentially means that the neuron is being trained to fire for certain sets of inputs and not fire for other sets of inputs. In the using mode, when a known input or input on which the neuron has been trained on is fed to the perceptron, the associated output is obtained. The using mode always occurs after the training mode. Various activation functions are used in a perceptron, they include step function, sigmoid function, hyperbolic-tangent function (TANH), RELU (Rectified Linear Unit) and sign function. A sigmoid function gives a smooth gradient and prevents jumps in output values. A step function cannot give multi-value outputs. TANH function is thought to be a zero-centred function; this means that it is sufficiently easier to model strongly negative, neutral and strongly positive values. The RELU function is computationally extremely efficient. Weights play an important role in the working of ANN; the perceptron can prioritize certain factors over others by the use of weights. Multilayer perceptron is otherwise known as the artificial neural network. A neural network can be thought as a combination of perceptrons which are connected in different ways and operating on different activation functions. There are three main layers in an artificial neural network: input layer, output layer and hidden layer. The number of hidden layers depends on the application; it can be different for different applications. The most common method by which the network can be trained is known as backpropagation. In this method, once the weighted sum of inputs is passed through the activation function, propagation is done backward and the weights are updated to reduce errors and, hence a more desirable output is obtained [17] (Fig. 14).

4.5.1 ANN in CDSSs

In CDSSs, ANN can be used to study the patient data, and these data are then mapped to symptoms, and hence, a possible diagnosis can be obtained. In order to achieve this functionality of neural network, the network must first undergo the training mode. In the training mode, the network is fed with a vast amount of clinical data, which include symptoms, signs, diagnosis, prognosis, medication, etc. The network then analyses these inputs and gives a possible outcome. The outcome is then correlated with actual clinical results, and by using the backpropagation method, the weights are adjusted to match with the actual clinical outcome. ANN can proceed even with incomplete data; the network makes educated guesses about the possible outcome, and by comparing with the actual clinical results, necessary modifications can be made. ANN is particularly advantageous as it eliminates the need for manual prescriptions and handwritten records. With proper training, the neural network can provide with accurate diagnosis and can help in the early detection of certain fatal diseases such as cancer. The main disadvantage of artificial neural network is that it is not cost effective, as a lot of money and time are required for the training of a neural network. The ANN recognizes patterns using the patient's data, and this ability of the neural network enables the ANN to be focused on a specific disease such as myocardial infarction, commonly known as heart attack [18]. The first application of ANNs in medical diagnostics came about in the late 1980s in the work by Szolovits et al. [17]; since then a number of different studies have come about in this field. ANNs have been used in the diagnosis of a variety of conditions such as colorectal cancer [19], pancreatic disease [20], early diabetes [21] and colon cancer [22]. ANN has also been widely used in the field of cardiology [23] and paediatrics [24]. Artificial neural networks have recently been used in the search for biomarkers [25]. Cancer and diabetes are the most common diseases found in the world population today. A study by the World Health Organization (WHO) reveals that around 30 million people of various ages and breeds suffer from various forms of diabetes 1. Due to its high prevalence, the clinical data obtained from the patients suffering from these diseases are vast. This large amount of clinical data provides an input to the neural network, and hence the properties of neural networks can be exploited to help in the early detection of such devastating diseases.

4.5.2   Applications in Clinical Studies

Cardiovascular Diseases
Cardiovascular diseases are a major health concern for majority of the world population. Cardiovascular disease is a disease class that involves the heart and blood vessels· A study by the American Heart Association shows that an estimated 17.3 million people lose their lives as a result of this disease per year, particularly due to heart attack, stroke, pulmonary heart disease, coronary heart disease, etc. These numbers are highly disturbing, and hence the need for an effective diagnostic tool that can detect the early onset of this disease became a requirement. A large amount of people suffering from such diseases generated a large amount of clinical data ranging from signs, symptoms, prognosis, diagnosis, medical imaging, etc. These data paved way for artificial neural networks to be used as diagnostic tools. For ANN to be used, data preparation is the foremost important step. Data which are obtained from the patient are categorized into various attributes, such as sex, age, weight, blood sugar and cholesterol level. Appropriate weights are assigned to each

of these attributes. The neural network is trained with these various attributes for effective heart disease prediction. The disadvantage is that ANN cannot accurately predict the type of heart disease, and hence it cannot be relied on for further diagnosis and medication.

Diabetes

Diabetes is another common disease found in the world population. It is a hetero-geneous group of multifactorial, polygenic syndromes characterized by elevated fasting blood glucose, or absolute absence of insulin. Diabetes is the major cause of adult blindness, renal failure, nerve damage, heart attack and a plethora of other conditions. Here again, the first step is data preparation. The data are collected from patients and are divided into attributes such as age, sex, fasting blood sugar and random blood sugar. Each attribute is assigned a particular weight. The data were collected periodically and were assessed for blood sugar levels. The set of input along with the respective weights is fed to the multi-layer perceptron, and the network is trained to predict the blood sugar level. This way patients need not make continuous visits to the hospital for blood tests, checking diastolic and systolic blood pressure, urine test, etc. To obtain more accurate results, the weights are modified such that it matches with the clinical result. Hence, using ANN can help in the early detection of diabetes, and it also provides a more accurate results, which paves way for better medical treatment and improved clinical management [21].

Cancer

A survey by the World Health Organization (WHO) revealed that nearly 16% of the world population die due to cancer and about 70% of these deaths occur in low- or middle-income countries. Worldwide only 14% of the people receive proper care. Cancer is a disease which has a better chance of being cured if detected early. Hence, highly accurate diagnostic and early detection systems are becoming a necessity. Artificial neural networks are trained so that they can identify the presence of cancer and the type of cancer if present. ANN is fed with inputs which include age, sex, cholesterol level and lifestyle attributes such as cigarette smoking and alcohol consumption. With appropriate training the network will be able to detect cancer and possibly predict which type it is. With more advances in technology, neural networks can now determine the type of brain tumour with the help of MRI images. Hence, ANN qualifies as an excellent diagnostic tool for the early detection of cancer and can enhance the quality of medical treatment [22].

Figure 15 provides an overview of the working of artificial neural networks in medical diagnosis. In the first step, a large amount of clinical data are fed to the system. The next step is a crucial step wherein certain features or attributes are selected; these attributes play in an important role as they determine what type of outcome will be obtained. The ANN is then trained and then verified. In the verification process, all the data that are redundant and that are capable of producing a negative result referred to as outliners are eliminated. The presence of outliners in the input can affect the training process and hence give rise to poorer results. Once all the outliners are removed, the database is said to be verified. This ends the

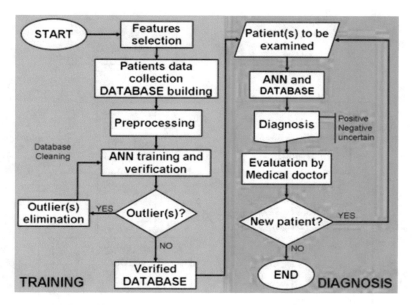

Fig. 15 Artificial neural networks in medical diagnosis [26]

training process. In the next step, the network is used for the diagnosis of the patient. In this step, the patients are examined and the required clinical data are obtained. These data are then fed as input the two multi-layer perceptron; the ANN takes all the inputs it has been trained for and correlates them with the desired result. The output is the diagnosis which can be positive, negative or uncertain. The diagnosis is then compared with the actual clinical diagnosis by a medical doctor and the course of medical treatment is decided. The neural network then prepares for new input from a new patient. This cycle continues. If more accurate data are fed to the ANN during training, a more accurate diagnosis can be obtained in the using mode [26].

4.6 Hybrid Systems

The efficiency and accurateness of solving any diagnosis in a rule-based reasoning is insufficient to pinpoint the different influences and causes. Hence, in order to enhance and optimize this simple structure, we combine the case-based reasoning and rule-based reasoning techniques to develop an effective model in the medical domain for controlling and handling knowledge-based structure, which can produce better results than the working of individual units separately.

The architecture of the proposed clinical decision support system is shown in Fig. 16. The system consists of the user interface, knowledge/case base, reasoning

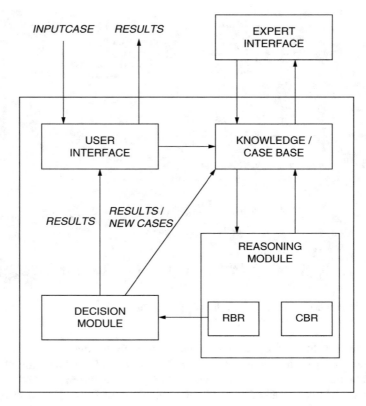

Fig. 16 CDSS architecture [27]

module, decision module and the expert interface. The user interface is responsible for the input of different cases or symptoms, which is then transmitted to the reasoning module via knowledge/case base. Knowledge/case base module is responsible to check whether the given input is distinct and unique from all the cases present in its database, and if unique, it adds it to its database. From the above architecture, it is confirmed that the numerous inference methods include rule-based reasoning, fuzzy logic, artificial neural networks and Bayesian networks among other mechanisms. Since rule-based reasoning comprises of a specific set of knowledge base using "if-then" logic, clinical-based reasoning enhances with the four R's technique as shown below [27] (Fig. 17).

The four R's include:

1. Retrieve: Involving retrieving different cases from the memory.
2. Revise: Forming associativity and mapping the present cases with any previous cases.
3. Reuse: Verify if the solution obtained is a distinct and new solution.
4. Remember: Only new distinct solutions are stored thereby reducing redundancy.

Fig. 17 The four R's in clinical-based reasoning [27]

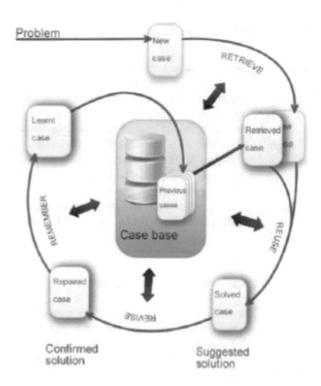

During the search operation, the system builds relationships between similar cases once retrieved. There exists a threshold which defines a specific case to old or new deepening on its associated value assigned after the operation. Due to major differences, the revised solution is generated by the adaptation process. This hybrid system can be used in the diagnoses of variety of diseases like lung diseases, diabetes and heart diseases. For an effective reasoning approach, the different methodologies can be combined. Case-based reasoning is mainly used due to its simplified empirical clinical cases. This also has certain drawbacks that include the measurement of cases and retrieval process but can be substantially improved with the help of rule-based reasoning. In these cases, the rule-based reasoning only deals with firing the basic rules and the major decision-making system is handled by the case-based reasoning. Hence, the rule-based reasoning is extremely effective in dealing with explicit knowledge and clinical-based reasoning works effectively in implicit knowledge base [27].

4.7   Genetic Algorithm

Genetic algorithms fall under the category of non-knowledge-based CDSS. Genetic algorithms are highly sophisticated algorithms based on natural selection and genetics. They are randomized optimization algorithms. Genetic algorithm was one of the first bio-inspired computational methods. Genetic algorithms are based on Charles Darwin's theory of natural selection. Just as organisms in the environment constantly adapt and adjust to changing conditions, genetic algorithms also adapt to optimize results. As with Darwin's theory of "survival of the fittest," genetic algorithms generally begin by attempting to solve a problem through the use of randomly generated solutions. Genetic algorithms continuously modify the population, and with progressive step, the algorithm chooses random parents to produce an offspring. As the generations further, the population evolves and provides an optimal solution.

4.7.1   Structure of Genetic Algorithm

Figures 18 and 19 describe the structural aspects of a genetic algorithm. The first step in a genetic algorithm is to randomly generate a population which has the ability to give rise to a solution. The population is represented by chromosomes. The chromosomes are character strings and are essentially encoded solutions to a particular problem. In the next step, the fitness of each chromosome is evaluated. Depending on what solution is required, a specific fitness criterion is set. The fitness criterion is a fitness function also known as evaluation function. This function estimates the closeness of a given solution to the optimal solution. Once fitness is assigned, selection error must be evaluated. If there is high selection error, then the fitness is low. Hence, it is a necessity to reduce the selection error to obtain optimal solutions. The attributes with greater fitness level are more likely to be selected in the population. Different methods can be used for fitness assignment, but the most commonly used method is the rank method. The next step is the application of genetic operators. Here, the attributes that satisfy the fitness function are selected. The genetic operator known as the selection operator will make this selection. The selected attributes are parents which give rise to the next generation. The most commonly used selection method is the roulette wheel method. In this method, the

Fig. 18 Structure of a
genetic algorithm

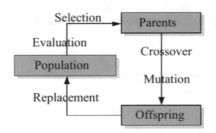

Fig. 19 The evolutionary
cycle

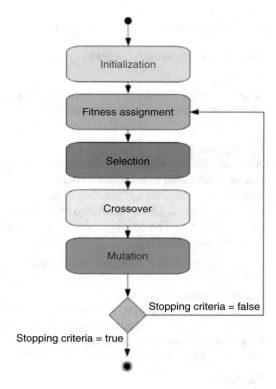

selected attributes are placed on a roulette wheel and are enclosed in areas that are proportional to their fitness. The next operator in action is the cross-over operator. This operator is involved in the recombination of the selected attributes to generate the new population or the next generation. The operators select two individuals at random and recombine them to generate four offspring; this process continues till the new population size matches the old one. Hence, at the end of the crossover, the population size remains constant. The first-generation offspring generated are very similar to their parents, and this results in loss of diversity. To solve this problem, the mutation operator comes into the picture. The mutation operator mutates some of the offspring at random by changing their values. This leads to a more diverse population. Each iteration in the cycle produces a new set of chromosomes. Figure 19 depicts the evolutionary cycle. Typically, genetic algorithms run from 50 to 1000 generations. At the end of any typical genetic algorithm run, there is at least one chromosome that is highly fit. The main advantage of using genetic algorithm is that it is faster and more efficient compared to other traditional methods. It provides us with a set of good solutions and not just a single solution. Genetic algorithms have the ability to manage a data set which is encompassed with many features. The main disadvantage of using genetic algorithm is that it is very expensive in computational terms and a lot of time has to be invested in making the prediction model [28].

4.7.2 Applications in Clinical Studies

Genetic algorithms have been widely used in clinical studies.

Oncology

With an aim to provide non-invasive diagnosis for cervical cancer, genetic algorithms have been used. A number of studies revealed that genetic algorithms were successfully able to differentiate between a normal and dysplastic cervix [29].

Paediatrics

A cheap and non-invasive technique to monitor and assess foetal heart rate and uterine contraction is achieved through cardiotocography. In his study, Ocak [30] applied the principles of genetic algorithms to select the most optimal readings from cardiotocography. This enabled him to further optimize the support vector machine (SVM) classifier. The now-optimized system could classify foetal health conditions with 99% accuracy [30].

Autism is another common neuro-related condition that occurs in children. Latkowski and Osowski in their study used genetic algorithms to identify the genes that most frequently occurred in association with this disease.

The most common type of blood cancer or leukaemia occurring in children is acute lymphoblastic leukaemia (ALL). Various subtypes of ALL are known to occur frequently. In a study by Lin et al., genetic algorithms were used to select the genes that were required for the accurate classification of ALL [31].

Cardiology

Myocardial infarction is a leading cardiovascular condition. It is commonly referred to as heart attack. Formation of plaque is one of the major reasons for the incidence of this condition. If medical professionals could determine the properties of the plaque, a better medical diagnosis and treatment can be mapped out. Khalil et al. in his study used genetic algorithms to determine one of the properties, elasticity.

Major adverse cardiac event (MACE) can be predicted using genetic algorithms. Zhou et al. used genetic algorithms to predict the risk of MACE.

Q wave, R wave and S wave of the electrocardiogram combine to give what is known as the QRS complex (Fig. 20).

A complete analysis and understanding of the QRS complex are essential for reading and interpreting the ECG. Tu et al. employed genetic algorithms to detect these QRS complexes [32].

Endocrinology

Very low level of blood sugar level causes a condition known as hypoglycaemia. Hypoglycaemia is an indicator of a health problem. The symptoms of hypoglycaemia include anxiety, fever, shakiness and nausea among others. Hypoglycaemia can induce changes in electroencephalograms (EEGs). Nguyen et al. used genetic algorithms along with ANN to predict hypoglycaemia based on EEG signals [33].

Fig. 20 QRS complex [32]

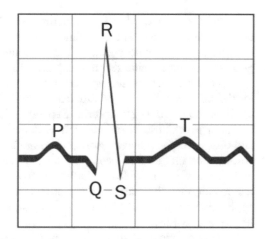

5   CDSS in Clinical Practice

5.1   *Virtual Psychiatrist*

In countries like India, mental health-care services are in their infancy stages due to a huge mental gap. India being a low- and middle-income (LAMI) country is deficient in resources and workforce pertaining to mental health care. The usual way to bridge this mental gap is to train physicians to identify different types of mental disorders and provide the respective care required. However, there are many roadblocks to solve these issues such as difficulty in training personnel and inadequate funding [34].

With the aim of solving these issues, a project to deliver psychiatric services to three remote sites with the help of digital information communication technology was initiated at the Department of Psychiatry at the Postgraduate Institute of Medical Education and Research (PGIMER), Chandigarh, India [35].

Hence, a model for digital mental health care was developed, and its potential for service delivery in LAMI countries was determined. The model was powered by an online fully automated clinical decision support system (CDSS). It has modules for diagnosis, treatment, management and follow-up, and it is usable by non-specialists with brief training and nominal supervision by psychiatrists [36].

Sites in Himachal Pradesh (HP), Uttarakhand (UK) and Jammu and Kashmir (JK), Chandigarh being the nodal site, were chosen for the assessment. The three sites have very large stretches of geographically difficult and inaccessible terrains, with a very low number of psychiatrists.

The project consists of two components:

1. Development and standardization of the diagnostic and management application (or CDSS)
2. Deployment of the digital application service and testing in real time.

A brief overview of the development of the CDSS application software has been provided below. These modules were based on conditional logic and clinical knowledge to generate specific diagnosis and quickly sort out patient data and generate patient-specific recommendations [37].

A set of mental illnesses were taken into consideration based on the *International Statistical Classification of Disease-10* and *Diagnostic and Statistical Manual of Mental Disorders-IV* criteria. Separate CDSS applications were created for common psychiatric disorders such as dementia, depression, alcohol dependence and obsessive-compulsive disorder. The modules developed contained an essential core section and also an additional history section. The additional history section entails many sub-modules including family history, stress factors, environmental factors, physical illnesses and many other factors affecting mental health. The CDSS application format could be read in English or Hindi, depending on the patient's lingual ability [38].

The diagnostic module had two types of questioning sub-modules, which are the "Rater's rules" and the "Decision Rules." Rater's rules stated how an interviewer should rate an item as present or absent, based on the objective of the question. This allowed to create a routine clinical environment. The screening module acts as the first checkpoint to be passed, and then based on the responses in the first round, specific questions are asked based on an inbuilt hierarchy. Altogether, there are three levels of hierarchy for a certain mental illness, and a certain checkpoint has to be crossed in order to reach the next level. Decision rules are based on the diagnostic thresholds set by the official classifications [39] (Fig. 21).

Finally, a summary profile of the diagnosis of the patient is generated. These profiles generated can be used by a general physician or a psychologist to provide a

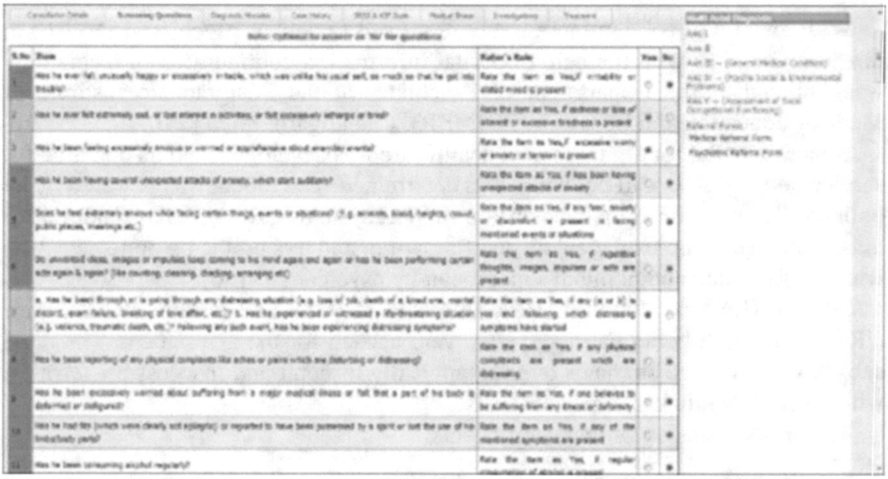

Fig. 21 Screenshot of the screening module of the clinical decision support system [36]

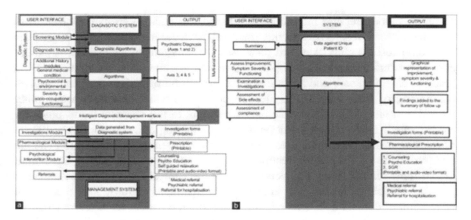

Fig. 22 (a) First consultation. (b) Flow of the system of the follow-up consultation [36]

better interpretation to the patient of what they are going through. Figure 22 shows the workflow of the CDSS and the way different decisions are made.

Finally, a follow-up module contains instructions on assessing improvement. It might also make changes to the treatment if required and records side effects or new symptoms if produced, and it also refers the patient for specialist care when required.

A majority of the patients found the CDSS application user friendly and were satisfied with the language used and the style in which the interview was conducted. The physicians and psychiatrists at the remote sites were satisfied with the choice of drugs and prescribed drug dosages. The counselling and relaxation training modules were useful and easy to understand by the patients or their caregivers. Hence, most of the patients were satisfied with the comprehensive assessment the CDSS application was able to provide [36].

5.2   Coronary Computed Tomography-Based Clinical Decision Support System

A coronary computed tomography-based CDSS has been designed to provide coronary decision support. This system uses a machine-learnt predictor to predict clinical decision for patients based on various input sources. This support may be provided to the physician prior to reviewing coronary CT data, thus helping with the decision to send patients to the catheterization laboratory. This is beneficial as a significant number of patients undergoing the procedure are reported to have no ischemia causing lesions [40].

The coronary CT data representing the patient's heart are obtained from the CT system. In addition to this, non-invasive patient data like patient history, demographics, blood pressure measurements, blood test data, molecular measurement information and results from other medical devices are obtained through tests and computerized clinical record databases. An anatomical evaluation of patient coronary arteries comprising of plaque size and volume, stenosis grading, lesion grading, calcium score and a physiological evaluation of patient coronary arteries comprising of IFR, WSS, FFR, etc. are performed using quantitative and other tools. From this, data values which serve as input to the machine-learnt predictor are extracted. The machine-learnt predictor, comprising of a cascade or parallel set of machine-learnt classifiers, uses trained machine-learning algorithms to generate a clinical decision for patient treatment based on the input data values. The output is transmitted in the form of a decision tree of the clinical decision predicted, other possible clinical decisions and recommended treatment options. This aids in the generation of clinical decisions regarding sending the patient for invasive stenting, invasive measurement, non-invasive tests, medicine prescription and discharge. This CDSS system can help physicians in charting a treatment path personalized for different individuals [40].

6 Drawbacks of CDSS

Drawbacks of CDSS mainly result from improper data storage and retrieval resulting in inaccurate datasets. The resistance to its adoption in clinical settings due to possible machine errors and the impersonal nature of CDSS is another major drawback.

Dependence on Centralized Data Repositories
Knowledge base is a critical architectural component of CDSS, which is dependent on a centralized repository of clinical data. This calls for the need of standardized representation of data. Repositories can be used to manage the storage and retrieval of these data. Lack of a well-managed and accurate knowledge base can impact the quality of clinical counsel offered by the CDSS. Good-quality repositories are required by data mining algorithms to be able to extract the relevant information necessary for clinical decision-making.

Large volumes of clinical datasets are obtained from content available on EHR, EMR and PHR. Corrupt databases can result from lack of standardized data capture methods. Corrupt databases provide an inaccurate representation of the patient population. It is, therefore, essential that data repositories and data capture methods are accurate to enable mining algorithms to extract patient-specific care recommendations [41].

Efficient Knowledge Management
The efficient assimilation of knowledge requires certain rules and guidelines. Knowledge management emphasizes on the articulation, capture and distribution of explicit and tacit knowledge in different formats. This is required to assimilate

existing knowledge to derive new knowledge. Creation of new knowledge is routinely observed in the medical field in the creation of new databases to define new disease and treatment methods. However, if the existing data on which the knowledge is based is of substandard quality, the outcome can be devastating in a clinical setting. Therefore, appropriate knowledge management is necessary to ensure patient-specific treatment recommendations. Information obtained from the CDSS must be analysed by health-care providers before decisions are made [41].

Improper Interpretation of Clinical Data

Data available in the CDSS are prone to misinterpretation and misrepresentation. This can arise from machine-related errors and can result in misdiagnosis. Further, automation biases generate errors that might not correlate with the practical knowledge of physicians. It is also noted that real data obtained from patient interviews and medical records are not appropriately organized and can alter the performance considerably [41].

Possible Source of Inconvenience to Clinicians

In actual clinical settings, it cannot be practically feasible for a clinician to consult a CDSS frequently during routine examinations; CDSS can sometimes increase the workload of physicians, and routine consultation with a system can alienate them from direct contact with their patients. Health-care providers with experiential knowledge are less likely to consider clinical decisions presented by a CDSS. Experienced physicians choose to rely on personal knowledge to devise a strategy for treatment rather than relying on existing clinical data. These attitudes can impede the acceptance and usage of CDSS in clinical practices. CDSS should therefore be customizable to suit the needs of the clinician [41].

7 Ethical and Legal Issues

Some of the ethical and legal issues related to CDSS have been summarized below.

Care Standard

The use of a CDSS might increase the risk of error. There are instances where despite accurate preliminary diagnosis made with the help of data collected from a patient, the treatment can turn out to be inaccurate, challenging the initial diagnosis. In such cases, a trained medical practitioner must use his expertise in deducing the appropriate treatment, something a system might not be capable of. Some patients might not be very comfortable with their diagnosis being determined by an informatics tool. Use of such tools will also result in the lack of personal interaction of the physician with the patient, which might result in inefficient data collection. All these challenges to the practical implementation of CDSS raise certain ethical questions and make clinicians reluctant to use the system to its full potential [18].

Appropriate Use
Appropriate use of a CDSS is important for ensuring good patient care. If a CDSS designed with an educational intent is depended upon for clinical decision support or if a system designed for modest decision support is used in a way that diagnosis by clinical experience is abandoned, the results can be disastrous. Therefore, the users of CDSS must have the appropriate qualifications and training to efficiently use the system [18].

Liability
The main legal issue associated with CDSS is the liability for misuse of the system to make or assist in medical decisions. In case a patient is injured due to defective diagnosis by the system, the clinician must be liable for negligence [18].

8 Conclusion and Future Scope

Clinical Decision Support Systems have provided a wealth of information to scientists and physicians to make patient-specific medical decisions using advanced computational approaches. CDSS helps in improving patient care by assisting a clinician from initial consultation to diagnosis and treatment. The future applications of this technology are numerous with its adoption in fields like pharmacology, pharmacogenomics, predictive medicine, antimicrobial treatment and pathology. This system has been used for providing antimicrobial treatment. This is achieved by interpreting patient data on the history of antibiotic medication administered, microbiology data and drug allergy information. CDSS has been shown to select the appropriate antimicrobial treatment regimens more frequently than physicians. The use of computational data assessment tools like CDSS in medicine has provided an opportunity for its utilization in antimicrobial and antibiotic resistance research [42]. It has been used in drug allergy checking and drug dosage. Future applications will be to optimize dosage selection using low-cost genomic analysis which would enable the prediction of drug metabolism based on the genetic makeup of patients. Although this might require a large data set, once established, a CDSS tailored to this purpose will permit patient-specific evidence-based treatment options [43]. It has also been used in the analysis of pathology reports. Advances in EHR databases and histopathological techniques have allowed for acquisition of large amount of patient data. Such CDSSs have been used to pool cancer data and provide prognostic tools. They have also been used to predict probabilities of inheritance of specific mutations in cancers by analysing patient data and family history. This illustrates the immense potential of medical informatics to guide preventive medicine. If improved, these systems can be valuable tools in the treatment of conditions like cancer which requires multidimensional approaches [43].

The significance of CDSS is especially evident in underdeveloped and developing countries which report higher incidences of nutritional disorders and outbreaks. A well-managed CDSS can equip such countries to handle such emergencies by providing real-time patient history. This can have a major impact in terms of improvement of general health of the population. The combination of medicine and informatics has resulted in numerous important advancements in medical treatment. The adoption and implementation of CDSS can assist clinicians and benefit patients by decreasing cost and increasing the standard of health care. CDSS thus serves as an effective system to monitor and improve health care globally and has the potential to revolutionize the future of health care.

Conflict of Interest The authors have no conflict of interest. There was also no problem related to funding. All authors have contributed equally with their valuable comments which made the manuscript to this form.

References

1. Ledley, R., & Lusted, B. (1959). Reasoning foundations of medical diagnosis. *Science, 130*(3366), 9–21. https://doi.org/10.1126/science.130.3366.9.
2. Quintana, Y., & Safran, C. (2016). Global Health informatics–an overview. In H. d. F. Marin, E. Massad, M. A. Gutierrez, R. J. Rodrigues, & D. Sigulem (Eds.), *Global Health informatics: How information technology can change our lives in a globalized world* (pp. 1–13). Oxford: Academic Press. isbn:978-0-12-804591-6.
3. Kaushal, R., Shojania, K. G., & Bates, D. W. (2003, Jun 23). Effects of computerized physician order entry and clinical decision support systems on medication safety: A systematic review. *Archives of Internal Medicine, 163*(12), 1409–1416.
4. Jao, C. S., & Hier, D. B. (2010, Jan 1). Clinical decision support systems: An effective pathway to reduce medical errors and improve patient safety. In C. S. Jao (Ed.), *Decision support systems*. IntechOpen. https://doi.org/10.5772/39469. Available from: https://www.intechopen.com/books/decision-support-systems/clinical-decision-support-systems-an-effective-pathway-to-reduce-medical-errors-and-improve-patient.
5. Gorgulu, O., & Akilli, A. (2016, Oct 1). Use of fuzzy logic based decision support systems in medicine. *Studies on Ethno-Medicine, 10*(4), 393–403.
6. Wadgaonkar, J., & Bhole, K. (2016). Fuzzy logic based decision support system. In *2016 1st India international conference on information processing (IICIP)* (pp. 1–4). Delhi. https://doi.org/10.1109/IICIP.2016.7975310.
7. Samad-Soltani, T., Ghanei, M., & Langarizadeh, M. (2015, Jun). Development of a fuzzy decision support system to determine the severity of obstructive pulmonary in chemical injured victims. *Acta Informatica Medica, 23*(3), 138.
8. Anooj, P. K. (2012, Jan 1). Clinical decision support system: Risk level prediction of heart disease using weighted fuzzy rules. *Journal of King Saud University-Computer and Information Sciences, 24*(1), 27–40.
9. Nee, O., & Hein, A. (2010, Mar 1). Clinical decision support with guidelines and Bayesian networks. In G. Devlin (Ed.), *Decision support systems advances*. IntechOpen. https://doi.org/10.5772/39394. Available from: https://www.intechopen.com/books/decision-support-systems-advances-in/clinical-decision-support-with-guidelines-and-bayesian-network.
10. Cypko, M. A., & Stoehr, M. (2019, Feb). Digital patient models based on Bayesian networks for clinical treatment decision Support. *Minimally Invasive Therapy & Allied Technologies, 26*, 1–5.
11. Seixas, F. L., Zadrozny, B., Laks, J., Conci, A., & Saade, D. C. (2014, Aug 1). A Bayesian network decision model for supporting the diagnosis of dementia, Alzheimer's disease and mild cognitive impairment. *Computers in Biology and Medicine, 51*, 140–158.

12. Sesen, M. B., Nicholson, A. E., Banares-Alcantara, R., Kadir, T., & Brady, M. (2013, Dec 6). Bayesian networks for clinical decision support in lung cancer care. *PLoS One, 8*(12), e82349.
13. Park, E., Chang, H. J., & Nam, H. S. (2018). A Bayesian network model for predicting post-stroke outcomes with available risk factors. *Frontiers in Neurology, 9.*
14. Rahaman, S., Islam, M. M., & Hossain, M. S. (2014). A belief rule based clinical decision support system framework. In *2014 17th international conference on computer and information technology (ICCIT)* (pp. 165–169).
15. Karim, H., Hosseini, M. R., Zandesh, Z., et al. (2019). A unique framework for the Persian clinical guidelines: Addressing an evidence-based CDSS development need. *BMJ Evidence-Based Medicine.* https://doi.org/10.1136/bmjebm-2019-111187.
16. Anderson, J. A., & Willson, P. (2008, May 1). Clinical decision support systems in nursing: Synthesis of the science for evidence-based practice. *CIN: Computers, Informatics, Nursing, 26*(3), 151–158.
17. Szolovits, P., Patil, R. S., & Schwartz, W. B. (1988, Jan 1). Artificial intelligence in medical diagnosis. *Annals of Internal Medicine, 108*(1), 80–87.
18. Berner, E. S. (2007). *Clinical decision support systems.* New York: Springer Science+ Business Media, LLC.
19. Spelt, L., Andersson, B., Nilsson, J., & Andersson, R. (2012, Jan 1). Prognostic models for outcome following liver resection for colorectal cancer metastases: A systematic review. *European Journal of Surgical Oncology (EJSO), 38*(1), 16–24.
20. Bartosch-Härlid, A., Andersson, B., Aho, U., Nilsson, J., & Andersson, R. (2008, Jul). Artificial neural networks in pancreatic disease. *British Journal of Surgery: Incorporating European Journal of Surgery and Swiss Surgery, 95*(7), 817–826.
21. Shankaracharya, D. O., Samanta, S., & Vidyarthi, A. S. (2010). Computational intelligence in early diabetes diagnosis: A review. *The Review of Diabetic Studies: RDS, 7*(4), 252.
22. Ahmed, F. E. (2005, Dec). Artificial neural networks for diagnosis and survival prediction in colon cancer. *Molecular Cancer, 4*(1), 29.
23. Turkoglu, I., Arslan, A., & Ilkay, E. (2003, Jul 1). An intelligent system for diagnosis of the heart valve diseases with wavelet packet neural networks. *Computers in Biology and Medicine, 33*(4), 319–331.
24. Bhatikar, S. R., DeGroff, C., & Mahajan, R. L. (2005, Mar 1). A classifier based on the artificial neural network approach for cardiologic auscultation in pediatrics. *Artificial Intelligence in Medicine, 33*(3), 251–260.
25. Bradley, B. P. (2012, Apr 1). Finding biomarkers is getting easier. *Ecotoxicology, 21*(3), 631–636.
26. Amato, F., López, A., Peña-Méndez, E. M., Vaňhara, P., Hampl, A., & Havel, J. (2013). Artificial neural networks in medical diagnosis. *Journal of Applied Biomedicine, 11*, 47–58.
27. Kong, G., Xu, D.-L., & Yang, J.-B. (2008). Clinical decision support systems: A review on knowledge representation and inference under uncertainties. *International Journal of Computational Intelligence Systems., 1*, 159–167. https://doi.org/10.1080/18756891.2008.9727613.
28. Tafani, D. (2012). *Analytic modelling and resource dimensioning of optical burst switched networks.* https://doi.org/10.13140/RG.2.2.12436.09605.
29. Zhou, X., Wang, H., Wang, J., Wang, Y., Hoehn, G., Azok, J., Brennan, M. L., Hazen, S. L., Li, K., Chang, S. F., & Wong, S. T. (2009, Apr). Identification of biomarkers for risk stratification of cardiovascular events using genetic algorithm with recursive local floating search. *Proteomics, 9*(8), 2286–2294.
30. Ocak, H. (2013, Apr 1). A medical decision support system based on support vector machines and the genetic algorithm for the evaluation of fetal well-being. *Journal of Medical Systems, 37*(2), 9913.
31. Lin, T. C., Liu, R. S., Chao, Y. T., & Chen, S. Y. (2013, Apr 10). Classifying subtypes of acute lymphoblastic leukemia using silhouette statistics and genetic algorithms. *Gene, 518*(1), 159–163.

32. Tu, C., Zeng, Y., & Yang, X. (2005, Jan 1). A new approach to detect QRS complexes based on a histogram and genetic algorithm. *Journal of Medical Engineering & Technology, 29*(4), 176–180.

33. Nguyen, L. B., Nguyen, A. V., Ling, S. H., & Nguyen, H. T. (2013). Combining genetic algorithm and Levenberg-Marquardt algorithm in training neural network for hypoglycemia detection using EEG signals. In *2013 35th annual international conference of the IEEE engineering in medicine and biology society (EMBC)* (pp. 5386–5389). Osaka. https://doi.org/10.1109/EMBC.2013.6610766.

34. Murthy, R. S. (2014, Aug 5). Mental health initiatives in India (1947-2010). *National Medical Journal of India, 24*(2), 98–107. PMID-21668056.

35. Malhotra, S., Chakrabarti, S., Shah, R., Gupta, A., Mehta, A., Nithya, B., Kumar, V., & Sharma, M. (2014, Dec). Development of a novel diagnostic system for a telepsychiatric application: A pilot validation study. *BMC Research Notes, 7*(1), 508.

36. Malhotra, S., Chakrabarti, S., & Shah, R. (2019, Jan). A model for digital mental healthcare: Its usefulness and potential for service delivery in low- and middle-income countries. *Indian Journal of Psychiatry, 61*(1), 27.

37. Cresswell, K., Majeed, A., Bates, D. W., & Sheikh, A. (2013, Mar 22). Computerized decision support systems for healthcare professionals: An interpretative review. *Journal of Innovation in Health Informatics., 20*(2), 115–128.

38. World Health Organization. (1992). *The ICD-10 classification of mental and behavioural disorders. Clinical descriptions and diagnostic guidelines.* Geneva: World Health Organization. Accessed December 10, 2014. *The Journal of Clinical Psychiatry* 1997;58(suppl 8):27–34.

39. Malhotra, S., Chakrabarti, S., Shah, R., Sharma, M., Sharma, K. P., Malhotra, A., Upadhyaya, S. K., Margoob, M. A., Maqbool, D., & Jassal, G. D. (2017, Aug). Telepsychiatry clinical decision support system used by non-psychiatrists in remote areas: Validity & reliability of diagnostic module. *The Indian Journal of Medical Research, 146*(2), 196.

40. Sharma P, Itu LM, Flohr T, Comaniciu D, inventors; Siemens Healthcare GmbH, assignee. (2019, Jan 8). *Coronary computed tomography clinical decision support system.* United States patent application US 10/176,896.

41. Bonney, W. (2011, Sep 6). Impacts and risks of adopting clinical decision support systems. In C. Jao (Ed.), *Efficient decision support systems – Practice and challenges in biomedical related domain.* IntechOpen. https://doi.org/10.5772/16265. Available from: https://www.intechopen.com/books/efficient-decision-support-systems-practice-and-challenges-in-biomedical-related-domain/impacts-and-risks-of-adopting-clinical-decision-support-systems.

42. Laupland, K. B., & Valiquette, L. (2013). Outpatient parenteral antimicrobial therapy. The Canadian journal of infectious diseases & medical microbiology. *Journal canadien des maladies infectieuses et de la microbiologie medicale, 24*(1), 9–11. https://doi.org/10.1155/2013/205910.

43. Castaneda, C., Nalley, K., Mannion, C., Bhattacharyya, P., Blake, P., Pecora, A., Goy, A., & Suh, K. S. (2015, Dec). Clinical decision support systems for improving diagnostic accuracy and achieving precision medicine. *Journal of Clinical Bioinformatics, 5*(1), 4.

Part II
Natural Language Processing

Dynamic Mode Decomposition and Its Application in Various Domains: An Overview

S. Akshay, K. P. Soman, Neethu Mohan, and S. Sachin Kumar

1 Introduction

Dynamic mode decomposition (DMD) is a data-driven, matrix decomposition technique developed using linear Koopman operator concept [1]. The key feature of DMD algorithm is its ability to extract both spatial and temporal patterns of the data where existing methods are restricted to either of the patterns [2]. DMD algorithm found its application in a variety of domain-specific applications, such as fluid flow analysis, neural data analysis, load forecasting, parameter estimation, and image processing. In all these applications, DMD identifies the underlying dynamics of the associated system through measured data [3]. This peculiar ability of DMD makes it a suitable choice for different tasks.

This chapter gives an overview of DMD algorithm and its application in various fields as an emerging data-driven algorithm. In this chapter, the capability of DMD algorithm for complex data analysis is well explored with numerous examples. To show the effectiveness and potential of DMD, the examples are selected from various disciplines with distinct applications. The rest of the chapter is organized as follows. Section 2 gives the theoretical explanations and mathematical descriptions of the DMD algorithm in detail. Section 3 discusses the applications of DMD in different disciplines of science and engineering. The applications considered in this chapter are harmonics monitoring and parameter estimation in smart grid, complex flow analysis in fluid dynamics, short-term forecasting of electric loads, and image saliency detection.

S. Akshay · K. P. Soman · N. Mohan (✉) · S. Sachin Kumar
Center for Computational Engineering & Networking (CEN), Amrita School of Engineering,
Amrita Vishwa Vidyapeetham, Coimbatore, India

© Springer Nature Switzerland AG 2021
R. Kumar, S. Paiva (eds.), *Applications in Ubiquitous Computing*, EAI/Springer
Innovations in Communication and Computing,
https://doi.org/10.1007/978-3-030-35280-6_6

121

2   Dynamic Mode Decomposition Algorithm

The spatiotemporal data analysis tool known as dynamic mode decomposition (DMD) is built based on the concepts of singular value decomposition (SVD) [4]. DMD has revolutionized the fluid dynamics community due to its peculiar behavior to approximate the underlying dynamical characteristics of the system.

The DMD algorithm is closely related with proper orthogonal decomposition (POD), widely used in structural engineering and fluid dynamics. The POD is basically SVD. The data measurements for POD and DMD are assumed to be spatial data arranged in the form of a matrix. Mathematically speaking, the data we are capturing are a function over a rectangular field. The function value is assumed to be varying over time. To study its dynamics, snapshots are taken at regular interval of time. Figure 1 indicates the visualization of the data as snapshots over time.

The time-indexed vector data are arranged in the form of columns in a matrix X as shown in Fig. 2. The time index is varied from 1 to m.

The X matrix is defined as follows:

$$X = \begin{pmatrix} | & | & & | \\ x_1 & x_2 & \dots & x_m \\ | & | & & | \end{pmatrix} \tag{1}$$

The SVD factorization of matrix X yields,

$$X = U \Sigma V^T \tag{2}$$

Here, the data are assumed to be the sum of several independent spatial structures evolving over time. This interpretation is possible if we split the matrix as a sum

Fig. 1 Illustration of the data as snapshots

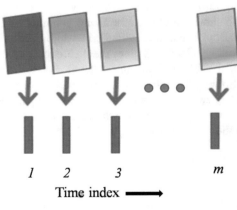

Time index ⟶

Fig. 2 Illustration of the data arranged as the columns of matrix X

Fig. 3 Illustration of two matrices U and ΣV^T

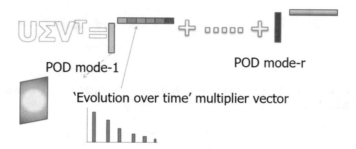

Fig. 4 Picturization of POD modes and its evolution over time

of several rank one matrices. For this, we consider the following two matrices as shown in Fig. 3.

It is possible to express X as outer product of vectors from these two matrices. It is defined as follows:

$$X = u_1\sigma_1 v_1^T + u_2\sigma_2 v_2^T + \ldots\ldots + u_m\sigma_m v_m^T \tag{3}$$

The columns of U are called POD modes and are pairwise and mutually orthogonal. Corresponding to each column vector u_k, there is a scaled row vector $\sigma_k v_k^T$ which specify the evolution of u_k over time indices from 1 to m and beyond. u_k in turn may be reshaped into the original matrix form. So, the term $u_k\sigma_k v_k^T$ may be visualized as a matrix of values (structural patterns in data) evolving over time. This is picturized in Fig. 4.

Upon comparing with POD, DMD is a much more powerful concept, and it assumes that the evolution of the function over the rectangular field is affected by the mapping of a constant matrix A. This concept is illustrated through Fig. 5.

DMD assumes that this A matrix captures the system's inherent dynamics, and thus, DMD objective is to find the A or equivalently its dominant eigenvalues and eigenvectors [5]. The size of the matrix is decided by the size of the column vector in X data matrix. If the column size is 1000×1, then size of the matrix A is 1000×1000. One assumption used here is that this matrix A is of low rank, and hence, the sequence of vectors $x_1, x_2, x_3, \ldots, x_k, \ldots, x_m$ finally becomes a linearly dependent set. That is, vector x_m becomes linearly dependent on previous vectors.

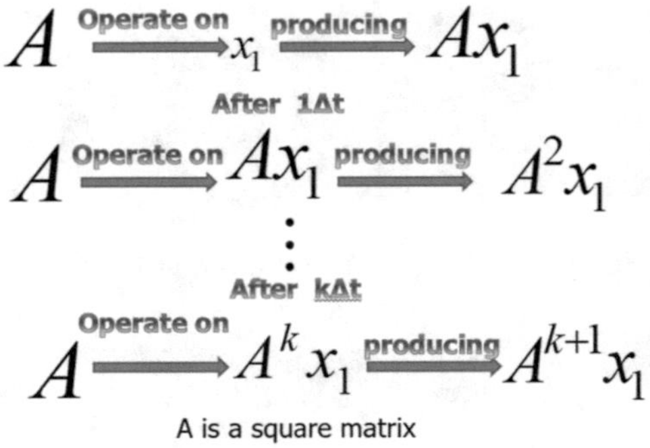

Fig. 5 Visualization of data mapping using A matrix

$$A \xrightarrow{\text{Operate on}} x_1 \xrightarrow{\text{producing}} Ax_1$$

After 1Δt

$$A \xrightarrow{\text{Operate on}} Ax_1 \xrightarrow{\text{producing}} A^2 x_1$$

$$\vdots$$

After kΔt

$$A \xrightarrow[\text{Operate on}]{} A^k x_1 \xrightarrow{\text{producing}} A^{k+1} x_1$$

A is a square matrix

Now, let us try to express data matrix X in terms of eigenvectors associated with matrix A as follows:

$$\begin{aligned}
A^k x_1 &= \Phi \Lambda^k \Phi^\dagger x_1 = \Phi \Lambda^k b \\
A^k x_1 &= \Phi \Lambda^k \Phi^\dagger x_1 = \Phi \Lambda^k b = x_{k+1}
\end{aligned} \tag{4}$$

where $\Phi^\dagger$ is pseudo inverse of Φ. Here, we assume matrix A of rank m. Hence, Φ is having only m columns. That exactly is the reason for taking pseudoinverse rather than inverse. The columns of Φ are called the DMD modes. Eq. (4) can be written as follows:

$$x_{k+1} = \Phi \begin{pmatrix} \lambda_1^k b_1 \\ \lambda_2^k b_2 \\ \vdots \\ \lambda_m^k b_m \end{pmatrix} = \begin{pmatrix} | & | & & | \\ \phi_1 & \phi_2 & \cdots & \phi_m \\ | & | & & | \end{pmatrix} \begin{pmatrix} \lambda_1^k b_1 \\ \lambda_2^k b_2 \\ \vdots \\ \lambda_m^k b_m \end{pmatrix} \tag{5}$$

For $k = 0$, Eq. (5) yields,

$$x_1 = \begin{pmatrix} | & | & & | \\ \phi_1 & \phi_2 & \cdots & \phi_m \\ | & | & & | \end{pmatrix} \begin{pmatrix} b_1 \\ b_2 \\ \vdots \\ b_m \end{pmatrix}$$

For $k = 1$, Eq. (5) yields,

$$x_2 = \begin{pmatrix} | & | & & | \\ \phi_1 & \phi_2 & \cdots & \phi_m \\ | & | & & | \end{pmatrix} \begin{pmatrix} \lambda_1^1 b_1 \\ \lambda_2^1 b_2 \\ \vdots \\ \lambda_m^1 b_m \end{pmatrix}$$

And finally, for $k = m - 1$, Eq. (5) yields,

$$
x_m = \left(\begin{matrix} | & | & & | \\ \phi_1 & \phi_2 & \cdots & \phi_m \\ | & | & & | \end{matrix} \right) \begin{pmatrix} \lambda_1^m b_1 \\ \lambda_2^m b_2 \\ \vdots \\ \lambda_m^m b_m \end{pmatrix}
$$

Now following the above relation, we can express data matrix X as the sum of m rank-1 matrices. Each such matrix can be thought of as time evolution of DMD mode.

$$
X = b_1 \phi_1 \left(1 \ \lambda_1^1 \ \cdots \ \lambda_1^m \right) + b_2 \phi_2 \left(1 \ \lambda_2^1 \ \cdots \ \lambda_2^m \right) + \cdots + b_m \phi_m \left(1 \ \lambda_m^1 \ \cdots \ \lambda_m^m \right)
$$

(6)

Now the remaining question is how to find the eigenvalues and eigenvectors of A in an efficient way. For answering this question, DMD forms two data matrices (X_1 and X_2) as defined below using the data measurements.

$$
X_1 = \left(\begin{matrix} | & | & & | \\ x_1 & x_2 & \cdots & x_{m-1} \\ | & | & & | \end{matrix} \right)
$$

(7)

$$
X_2 = \left(\begin{matrix} | & | & & | \\ x_2 & x_3 & \cdots & x_m \\ | & | & & | \end{matrix} \right)
$$

(8)

The relation between the data matrices, X_1 and X_2, is defined using the A matrix.

$$
X_2 = A X_1
$$

(9)

By taking the SVD of the data matrix, X_1

$$
X_1 = U \Sigma V^H
$$

(10)

By substituting Eq. (10) in Eq. (9) gives

$$
A = X_2 X_1^{\dagger} = \underbrace{X_2 V \Sigma^{\dagger} U^H}_{B} \Rightarrow AU = B
$$

(11)

However, for many practical applications, A will be a large dimension matrix and its eigendecomposition becomes a computational burden [6]. Hence, a rank reduced matrix, $\tilde{A}$, which shares the same nonzero eigenvalues of A is introduced to resolve the issue. For deriving the expression for $\tilde{A}$, consider expressing each data vector in terms of POD modes (the POD modes of data matrix X_1 are column vectors in U).

The data vector x_1 is expressed as the linear combinations of POD modes as follows:

$$U\tilde{x}_1 = x_1 \tag{11}$$

Equation (11) can also be thought as a projection of data vector x_1 on POD modes and is equivalent to $x_1^H U = \tilde{x}_1^H$. Here, the tuple size of $\tilde{x}_1$ is far less than that of x_1, since the numbers of POD modes are few (less than m). Following the above interpretation, the kth and $k + 1^{th}$ data vectors are expressed as follows:

$$x_k = U\tilde{x}_k \quad \text{and} \quad x_{k+1} = U\tilde{x}_{k+1} \tag{12}$$

By taking the concept explained in Fig. 5, it is possible to express x_{k+1} in terms of A as follows:

$$x_{k+1} = Ax_k \tag{13}$$

Now from Eq. (12) and Eq. (13),

$$U\tilde{x}_{k+1} = AU\tilde{x}_k \Rightarrow \tilde{x}_{k+1} = U^H AU\tilde{x}_k = \tilde{A}\tilde{x}_k \tag{14}$$

That is,

$$\tilde{A}\tilde{x}_k = \tilde{x}_{k+1} \tag{15}$$

The matrix $\tilde{A}$ defined in Eq. (15) represents the low-dimensional dynamical evolution. From this, we can easily infer original dynamics using the relation $x_k = U\tilde{x}_k$.

Now, by combining Eq. (11) and Eq. (15),

$$\tilde{A} = U^H AU \Rightarrow \tilde{A} = U^H B \tag{16}$$

The matrix $\tilde{A}$ is a very low-dimensional matrix compared to A and shares the same set of nonzero eigenvalues with A. The spectral decomposition of $\tilde{A}$ yields,

$$\tilde{A} = W\Lambda W^{-1} \tag{17}$$

Here, W denotes the eigenvector matrix and Λ denotes the eigenvalue matrix of $\tilde{A}$. The dynamic mode matrix, Φ, is obtained as follows:

$$\Phi = BW = X_2 V\Sigma^\dagger W \tag{18}$$

The columns of dynamic mode matrix, Φ, represent the eigenvectors of A.

3 Applications of DMD

This section explains the applications of DMD in different disciplines of science and engineering. Due to the increased availability of huge data measurements, the usage and application of data-driven algorithms like DMD are exponentially growing. An illustration of the usage of DMD in different disciplines is depicted through Fig. 1. As evident from Fig. 6, the DMD algorithm can be used for various domains, such as fluid dynamics, financial markets, text analytics, control, multimedia, smart grid, images, and medical sciences. This section describes a few important applications of DMD.

Smart grid: Due to the introduction of distributed generation systems, renewable energy sources, and flexible loads, electric grid is facing issues related to power quality, stability, reliability, control, and protection. In the recent past, DMD is widely being exploited for different smart grid applications, such as stability analysis and modal identification [7, 8]. Here, we are discussing the application of DMD for harmonics monitoring and parameter estimation in power grids [9]. The frequency information of power signals is possible to extract using the eigenvalues of dynamic mode matrix. In [9], the authors have developed a novel methodology for detecting the frequencies in power signals by employing shift-stacked data matrix concept and SVD hard thresholding. The stacking of multiple time-shifted copies of power signals to form the initial data matrices helps to overcome the limitation of DMD algorithm to extract the multiple frequency components. Further, the singular value-based hard thresholding eliminates the singular values that corresponds to noises and other uncertainties in power signals. The performance of DMD for detecting the multiple frequencies and associated amplitudes in power quality (PQ) signals are shown in Fig. 7. DMD has prominent results than various other data-adaptive algorithms, such as variational mode decomposition, empirical wavelet transforms, and synchro-squeezing transforms, for power signal analysis [10–12].

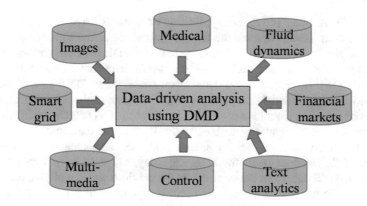

Fig. 6 Illustration of the data-driven analysis using DMD in different disciplines

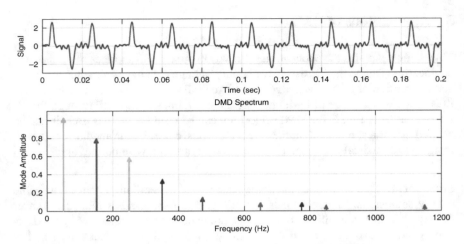

Fig. 7 Performance of the DMD for detecting the disturbance components in PQ signal

Fluid dynamics: Fluid flows are characterized by several temporal features and spatial coherent structures. Extraction of oscillatory modes and growth/decay rates of each mode is needed for the characterization of flow data. DMD algorithm is initially been proposed for turbulent fluid flow analysis [3]. DMD identifies the coherent structures of fluid flow and which can be used for extracting the oscillating modes and its growth/decay rates [13]. Understanding of these structures is important to characterize the dynamical behavior of fluid flow. The analysis by considering a dataset pertaining to fluid flow around a circular cylinder at Reynolds number $Re = 100$ using DMD is shown in Fig. 8. In this example, we are interested in computing the first $r = 15$ dominant DMD modes by considering the corresponding eigenvalues. The eigenvalues captured over the unit circle indicate the stable DMD modes, and these can be further used for finding the associated coherent structures in the data.

Forecasting: Another prominent application of DMD is forecasting/prediction of time-series data. DMD captures the features of the measured time-series data and is been utilized for prediction of future system state. DMD is mapped for financial forecasting in financial markets [14] and short-term load forecasting (STLF) in the power grid [15]. In [15], a novel data-driven strategy for STLF task is proposed using DMD. In this model, the ability of DMD to extract the meaningful, hidden tractable information from load series data is utilized for STLF. The main advantage of the model is the capability to handle the load series data that is affected by multiple factors, including time, day, seasons, climate, and socioeconomic activities. To forecast the load demand for a selected day, (1) data from two immediate previous days, (2) data from the same day in the previous week, and (3) data from the previous day in the previous week are used. Independent of heavy training stages, less amount of input data, and no parameter tuning are the key features of DMD-based STLF strategy. The DMD-based STLF model is adaptive to multiple seasonal

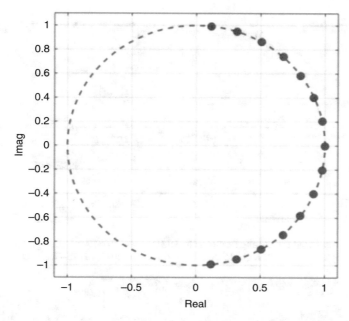

Fig. 8 The DMD eigenvalues correspond to 15 dominant modes in the absence of noise with *r = 15*

and cyclic patterns in load data and thus offers an improved generalization ability for load prediction of different seasons. On comparing with single stage and ensemble models, DMD-based model is less complex and offers fast estimation. Thus, this model can be used as an efficient tool for STLF in interconnected smart grid.

Figure 9 demonstrates the forecasting ability of DMD model for one-day ahead task. As clear from Fig. 9, the model predicts the day-ahead load more precisely than conventional autoregressive approaches.

Image analysis: The analytical property of DMD is been exploited for image processing applications. To effectively utilize DMD for static image processing applications, a dynamic representation of image data is needed. Here, we are discussing about saliency region detection and segmentation application of DMD [16]. In [16], the authors have proposed a novel idea to import dynamicity to static images by exploiting the color and luminance information. The full resolution salient maps are created in this way. Thus, this work utilized the analytical power of DMD for image saliency and segmentation application in color images. The developed saliency maps using DMD can be used for image classification, object detection, and image retrieval applications. The image saliency detection results of the DMD method is shown in Fig. 10. As evident from Fig. 10, the saliency maps created using DMD capture the saliency information in the images accurately.

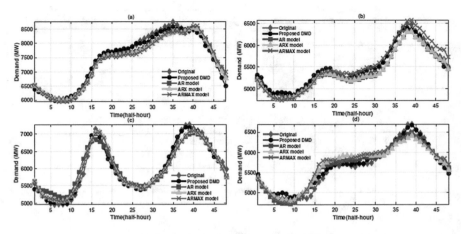

Fig. 9 One-day ahead forecasting results using DMD approach on Australian grid data. Forecasting results for (**a**) 10-Feb-2017, (**b**) 15-Apr-2017, (**c**) 27-Jul-2017, (**d**) 15-Oct-2017

Fig. 10 The results of DMD-based image saliency detection

4 Conclusion

The data-driven algorithms are widely explored to deal with the heterogeneous data available in different discipline of science, engineering, and medicine as it can extract more valuable information regarding the underlying system. The dynamic mode decomposition (DMD) is a leading data-driven algorithm that decomposes the complex systems into spatiotemporal coherent structures or modes. The modes can be used for identifying the latent dynamic characteristics of the underlying system. Originated in fluid dynamics community, DMD has gained its attention in different domains for various tasks, such as system analysis, forecasting, image analysis, video processing, and control systems. This chapter provides the details of DMD algorithm with mathematical explanations. The characteristics of DMD algorithm is explained with respect to POD and SVD, both are dominant techniques for data analysis. This chapter also covers the application of DMD in frequency estimation in smart grid, fluid flow analysis, short-term load forecasting, and image saliency detection tasks. Through the evaluation, it is concluded that DMD can be used as an emerging data-driven tool for various applications in different domains.

References

1. Schmid, P. J. (2010). Dynamic mode decomposition of numerical and experimental data. *Journal of Fluid Mechanics, 656,* 5–28.
2. Tu, J. H., Rowley, C. W., Luchtenburg, D. M., Brunton, S. L., & Kutz, J. N. (2014). On dynamic mode decomposition: Theory and applications. *Journal of Computational Dynamics, 1,* 391–421.
3. Kutz, J. N., Brunton, S. L., Brunton, B. W., & Proctor, J. L. (2016). Dynamic mode decomposition: Data-driven modeling of complex systems. *SIAM*.
4. Kumar, S. S., Manjusha, K., & Soman, K. P. (2014). Novel SVD based character recognition approach for Malayalam language script. In *Recent advances in intelligent informatics* (pp. 435–442). Cham: Springer.
5. Kutz, J. N., Fu, X., & Brunton, S. L. (2016). Multiresolution dynamic mode decomposition. *SIAM Journal on Applied Dynamical Systems, 15*(2), 713–735.
6. Kumar, S. S., Anand Kumar, M., Soman, K. P., & Poornachandran, P. (2020). Dynamic mode-based feature with random mapping for sentiment analysis. In *Intelligent systems, technologies and applications* (pp. 1–15). Singapore: Springer.
7. Barocio, E., Pal, B. C., Thornhill, N. F., & Messina, A. R. (2014). A dynamic mode decomposition framework for global power system oscillation analysis. *IEEE Transactions on Power Systems, 30*(6), 2902–2912.
8. Saldaña, A. E., Barocio, E., Messina, A. R., Ramos, J. J., Juan Segundo, R., & Tinajero, G. A. (2017). Monitoring harmonic distortion in microgrids using dynamic mode decomposition. In *2017 IEEE Power & Energy Society General Meeting* (pp. 1–5). IEEE.
9. Mohan, N., Soman, K. P., & Sachin, K. S. (2018). A data-driven approach for estimating power system frequency and amplitude using dynamic mode decomposition. In *2018 International Conference and Utility Exhibition on Green Energy for Sustainable Development (ICUE)* (pp. 1–9). IEEE.
10. Mohan, N., & Soman, K. P. (2018). Power system frequency and amplitude estimation using variational mode decomposition and chebfun approximation system. In *2018 Twenty Fourth National Conference on Communications (NCC)* (pp. 1–6). IEEE.
11. Prakash, L., Neethu Mohan, n., Kumar, S., & Soman, K. P. (2016). Accurate frequency estimation method based on basis approach and empirical wavelet transform. In *Proceedings of the Second International Conference on Computer and Communication Technologies* (pp. 801–809). New Delhi: Springer.
12. Soman, K. P., Sachin Kumar, S., Mohan, N., & Poornachandran, P. (2019). Modern methods for signal analysis and its applications. In *Recent advances in computational intelligence* (pp. 263–290). Cham: Springer.
13. Akshay, S., Gopalakrishnan, E. A., & Soman, K. P. (In Press). Investigating the effectiveness of DMD and its variants for complex data analysis. In *International Conference on Intelligent Computing and Control Systems – ICCS 2019*. IEEE.

14. Kuttichira, D. P., Gopalakrishnan, E. A., Menon, V. K., & Soman, K. P. (2017, September). Stock price prediction using dynamic mode decomposition. In *2017 International Conference on Advances in Computing, Communications and Informatics (ICACCI)* (pp. 55–60). IEEE.
15. Mohan, N., Soman, K. P., & Kumar, S. S. (2018). A data-driven strategy for short-term electric load forecasting using dynamic mode decomposition model. *Applied Energy, 232*, 229–244.
16. Sikha, O. K., Sachin Kumar, S., & Soman, K. P. (2018). Salient region detection and object segmentation in color images using dynamic mode decomposition. *Journal of Computational Science, 25*, 351–366.

Resolving Polysemy in Malayalam Verbs Using Context Similarity

S. N. Mohan Raj, S. Sachin Kumar, S. Rajendran, and K. P. Soman

1 Introduction

High degree of polysemy prevails in natural language, and so whatever utterance we come across is liable to be interpreted in multiple ways. But the high degree of ambiguity does not hamper our understanding of the concerned utterance. Mostly, the context invalidates the multiple interpretations and assigns a single interpretation to the given expression. It is the context which helps a native speaker to interpret an utterance or sentence correctly. Any automatic way of interpreting the sense of a lexical item expects the contexts to select or activate the correct sense out of the competitive senses of the concerned lexical items. All major word classes exhibit lexical ambiguity, and the contextual factors relevant for the concerned word resolve the meaning of the targeted word [20]. For example, adjectives are assigned meaning based on the head noun to which it is concatenated; similarly, the verb governing the noun or the modifier of the noun determines the meaning of the polysemous noun; the meaning of a verb may be determined by their argument structure or by the co-occurring elements of their syntactic frame ([24]: 215). This is illustrated below and the relevant senses are given in parentheses:

1. (a). *vEgatayuLLa kAr* "fast car" (the car that is or can be driven fast)
1. (b). *vEgatayuLLa jOlikkAr* "fast worker (one who works fast)"
2. (a). *patratte curuTTu* "role the newspaper" (physical object)
2. (b). *patratte paThikku* "read the newspaper" (content)
3. (a). *avar atinRe naSTatte uLkoNTu* "They accepted the loss" (pay)
3. (b). *avar A varttamAnam uLkoNTu* "They understood the information" (Learn)

S. N. Mohan Raj · S. Sachin Kumar (✉) · S. Rajendran · K. P. Soman
Center for Computational Engineering & Networking (CEN), Amrita School of Engineering,
Amrita Vishwa Vidyapeetham, Coimbatore, India
e-mail: kp_soman@amrita.edu

© Springer Nature Switzerland AG 2021
R. Kumar, S. Paiva (eds.), *Applications in Ubiquitous Computing*, EAI/Springer
Innovations in Communication and Computing,
https://doi.org/10.1007/978-3-030-35280-6_7

The lexical ambiguity of the verbs can be resolved by automatic means of finding sense distinctions using the semantics of the arguments of the targeted verbs. We can exploit the argument structure of a verb to interpret the correct sense of a verb. The corpus can give us the distribution of the verb with reference to its context. By means of distributional similarity, we will be able to select the correct sense of a verb. It is apparent that the resemblance of meaning between the words is revealed in resemblance of the contexts in which the words occur. It has been expressed in many ways as found in the "strong contextual hypothesis" of Miller and Charles [17] and in the well-known remark of Firth, "You shall know a word by the company it keeps" ([9]: 11).

It can be inferred that similar contexts in which a word occur reveal the similar sense of the word. This likelihood is exploited in resolving lexical ambiguity. However, making use of the notion of distributional similarity in computational applications faces problems. One of the main problems is that any kind of generalization based on distributional similarity relies on the identification of the sense in which a polysemous word occurs.

2 Types of Ambiguity

Ambiguity can emerge at any level of discourse – in words, in a sentence, or in a set of sentences. One can, thus, distinguish between referential, syntactical, and cross-textual ambiguities, depending on whether they occur in a single word, a sentence, or a set of sentences; a text. Pehar ([22]: 165–169) provided a definition and examples for each.

There are two types of ambiguity, lexical and structural. Lexical ambiguity is by far the more common. Everyday examples include nouns like *kAl* "quarter/leg," *aTa* "cake/incubation," *ara* "waist/half," and *maTi* "lap/laziness"; verbs like *Ota* "run," *aTi* "beat," and *eTu* "take"; and adjectives like *AzamuLLa* "deep," *uNakka* "dry," and *kaTTiyuLLLa* "hard." There are various tests for ambiguity. One test is having two unrelated antonyms, as with *kaTTiyuLLa* "hard," which has both *mrəduvAya* "soft" and *eLuppamuLLa* "easy" as opposites. Another is the conjunction reduction test. The test involves conjoining of two ambiguous sentences. The reduced sentence is zeugmatic. Consider the sentence, *avanu sarvakalASAlayil ninnum paTTam kiTTi; maRRonnu kaTayil ninnum kiTTi* "He received a degree from a university and another one from a shop." In the first part of the sentence *paTTam* refers to university degree and the second part it refers to "kite." Evidence that the word "*paTTam*" is ambiguous is provided by the anomaly of the "crossed interpretation" of the sentence, on which "*paTTam*" is used to refer to a university degree and "another one" to a kite.

The above examples of ambiguity are each a case of one word with more than one meaning. However, it is not always clear when we have only one word. The verb "desert" and the noun "dessert," which sound the same but, are spelled differently, count as distinct words (they are homonyms). So do the noun *muTTə* "knee" and the verb *muTTə* "knock," even though they not only sound the same but are spelled the same. These examples may be clear cases of homonymy. But in the case of noun *"kaTi"* and the verb *"kaTi"* or the postposition *mukaLil* "over" and the adverb *mukaLil* "over," the decision of homonymy is difficult to make. We have the problem of deciding whether they are pairs of homonyms or different forms of the same word. There is no general consensus on how to draw the line between cases of one ambiguous word and cases of two homonymous words. Perhaps, the difference is ultimately arbitrary.

Sometimes, one meaning of a word is derived from another. For example, the cognitive sense of *kANuka* "see" seems derived from its visual sense. The sense of *muRi* "break," in *avan maram muRi-ccu* "He cut the tree" is derived from its sense in *maram muRi-njnju* "The tree broke." Similarly, the transitive senses of *eri* "burn," *veTTu* "cut," and *poTTu* "burst" are derived from their intransitive senses. Now it could be argued that in each of these cases, the derived sense does not really qualify as a second meaning of the word but is actually the result of a lexical operation on the underived sense. This argument is plausible to the extent that the phenomenon is systematic and general, rather than peculiar to particular words. Lexical semantics has the task of identifying and characterizing such systematic phenomena. It is also concerned to explain the rich and subtle semantic behavior of common and highly flexible words like the verbs *cey* "do" and *iTu* "put" and the postpositions *uLLe* "inside," *mukaLil* "above," and *tAze* "below." Each of these words has uses which are so numerous yet so closely related that they are often described as "polysemous" rather than ambiguous.

Structural ambiguity occurs when a phrase or sentence has more than one underlying structure. For example, the phrases such as *kuLLarAya aNum peNNum* "short [men and women]," "[short man] and woman" and *veLLa marunnu kuppi* "white [medicine bottle]," "white medicine [bottle]" are structurally ambiguous. In the sentence, *kOzi tinnAn tayArAyi* "The chicken is ready to eat," that could be used to describe either a hungry chicken or a broiled chicken. It is arguable that the operative reading depends on whether or not the implicit subject of the infinitive clause "to eat" is tied anaphorically to the subject ("the chicken") of the main clause.

2.1 Lexical Ambiguity

This type of ambiguity as a rule connects to a homonym or a polysemous word presented in isolation (e.g., *aTa* "rice cake"/"blockade"/ "incubation"). Böhmerová ([4]: 28) presents the Table 1 as the typology of lexical ambiguity which is a partial representation of the numerous types of lexical ambiguity:

2.1.1 Lexical Ambiguity Due to Homonymy

If two entirely different words having same form have different meanings, the ambiguity arises due to homonymy. Homonymy is used in syntactic analysis to

Table 1 Types of lexical ambiguity

A. Language-inherent ambiguity (occurring in a particular language)
1. Onomatological (structural) ambiguity
2. Polysemy
3. Homonymy
4. Enantiosemy
B. Cross-linguistic ambiguity (occurring between or among languages)
1. Onomatological (structural) nonparallelism.
2. Lexical cross-linguistic nonparallelism (asymmetry)
3. Ambiguity due to synchronical motivational nontransparency
4. Ambiguity resulting from differing objective reality

refer to lexical items which have same form but differ in meaning. Homophones are a type of homonymy; it is used in semantic analysis to refer to words (i.e., lexemes) which have same pronunciation (e.g., *OTu* "tile"/"run"; *paTTam* "kite"/"degree"). The distinction between homographs and homophones is absent in Malayalam. When there is ambiguity between homonyms (whether nondeliberate or contrived, as in riddles and puns), a homonymic clash or conflict is said to have occurred. In semantic analysis, the theoretical distinction between homonymy and polysemy (one form with different meanings) provides a problem which has attracted a great deal of attention. Lyons ([15]: 550) identifies two kinds of lexical ambiguity, one of which depends on homonymy and the other on polysemy. He elaborately discuss about this problem giving more clarity to the concepts. He opines that the difference between homonymy and polysemy is easier to explain in general terms than it is to define in terms of objective and operationally satisfactory criteria. One of the criteria for distinguishing polysemy form homonymy is etymological information. Even etymological information may fail or unreliable sometimes. Lyons ([15]: 567) provides workable criteria based on distribution to solve the problem of homonymy and polysemy. In the following examples, the word *aTa* shows homonymy expressing contrastive sense; *aTa* can denote "rice cake" as well as "prop" or "keeping eggs for incubation."

4. (a). *avan aTa kazikkunnu* "He is eating rice cake"
4. (b). *avan vaNTiykkə aTa vaccu* "He put blockade for the cart"
4. (c). *avan muTTa aTa vaccu* "He kept the eggs for incubation"

As for machine translation is concerned both the homonymy and polysemy are treated alike as the aim is to finding out the meaning by context. The homographs belonging to different grammatical categories can be resolved as explained before. But they belong to same grammatical categories syntactic parsing may not be enough. One common approach is to assign semantic features such as *manushyan* "human," *strI* "female," and *drAvakam* "liquid" and to specify which features are compatible in the given syntactic constructions, via selection restrictions. For example, it might be specified that the verb *kuTi* "drink" have an "animate" subject.

But the expectation of animate subject for *kuTi* is challenged by the following type of usages where *kuTi* takes even inanimate subject.

5. *kARǝ orupATu indhanam kuTikkunnu* "The car is consuming lot of fuel"

There are difficulties in finding semantic features that can be used consistently and specifying the selection restriction for nouns and verbs based on these features. Even then, these are widely used in Machine translation system often in combination with case roles. But the semantic features cannot solve all the problems, even in situations for which they have been devised. For example, let us take the word *aTa*. As we have found out it is used in the senses of "rice cake," "keeping egg for incubation," and "prop." These three meanings can be differentiated explaining the relevant co-occurrence restrictions we find out in the following sentences in which *aTa* is used.

6. (a). *kuTTi aTa kazikkunnu* "The child is eating rice cake"
6. (b). *avan kOzhikkunjnju virikkAn vENTi irupatu muTTa aTa vaccu*

"He kept twenty eggs for incubation so as to hatch as chicken"

6. (c). *avan vaNTikku aTa vaccu* "He kept a prop for the cart"

The word *aTa* can be collocated with *kazikku* "eat" only in its eating sense; it can collocate with *vaykku* "keep" in the other two senses. *aTa* with "keeping for incubation" sense collocates with *muTTa* "egg" and with "prop" sense collocates with *vaNTi* "cart," thereby keep these two senses apart.

2.1.2 Lexical Ambiguity Due to Polysemy

If a word has two or more meanings, it can be said that the ambiguity is due to polysemy. Polysemy expresses extension of meaning. The polysemous words may express new meaning by metaphoric and metonymic extensions. For example, the word *SAkha* "branch" may denote branch of a tree as well as a branch of a bank. *naTa* can denote the action of walking as well as happening or functioning of something.

7. (a). *avan ennum rAvile sKULil naTannu pOkunnu*

"He goes to school daily by walking"

7. (b). *A stApanam nannAyiTTu naTakkunnuNTǝ*

"That organization is functioning well"

7. (c). *A tiyERRaRil sinima naTannukoNTirikkunnu*

"A cinema is running in the theatre"

OTu can denote human action of running as well functioning of mechanical devices.

8. (a). *avan vEgattil OTunnu* "He is running fast"
8. (b). cumar *ghaTikAram nannAyi OTunnu*.

"The wall clock is functioning well"

kaNNə may denote the eye of animate beings as well as the eye like spot in the coconut.

9. (a). *avan tanRe kaNNukaLe aTaccu* "He closed his eyes"
9. (b). *tEngaykkə mUnu kaNNukaL uNTə*

"there are three eye like spots in the coconut"

Similarly, many words denoting body parts show polysemous extension, thereby denoting ambiguity. For example, *vAy* can be a human mouth as well as the mouth of a bottle; *kAl* can be a human leg or leg of furniture. In the following example, the ambiguity is due to metonymic extension.

10. *grAmam ciriccu* "The people (of the village) laughed"

Here in this sentence, *grAmam* "people" is used as a metonymic extension of *grAmam* "village."

2.1.3   Polysemy, Homonymy, Delineating Senses

The polysemy–homonymy distinction is clear and unproblematic for the first sight. Homonyms are unrelated words that share the same spoken and written form, while a word that has two or more different, but related meanings is polysemous. The word *koTTu* is an example of polysemy, because it can refer to "a kind of percussion musical instrument," as well as "knock" and "beat." The similarity of their shape and function is reflected in their meaning; therefore, these two senses are said to be connected to the same, polysemous lexeme. Well-known examples for homonymy are *paTTam 1* "degree" and *paTTam 2* "kite." There could be dispute over the decision of homographic words as homonyms of polysemous words. In the case of *koTTu*, the senses are related form etymological or diachronic point of view. But *paTTam 1* and *paTTam 2* cannot be diachronically or etymologically related. The *paTTam* example shows that separating polysemy from homonymy may involve diachronic considerations. However, such a strategy should be aligned with the observation that speakers of a language are more or less unaware of the etymology of words, which also means that diachronically motivated polysemy–homonymy decisions lose their psycholinguistic relevance. Etymologically *naTa* "walk" and *naTa* "happen" and *naTa* "door step" are from the same source. They are given in the dictionary as three entries. On the other hand, when the history of the language is rejected as a clue, distinguishing polysemy from homonymy may turn out to be more than challenging.

Enumeration of senses in Natural Language Processing (NLP) applications is an accepted practice, too. NLP usually resort to what Lyons calls the "maximize homonymy" approach – by neglecting polysemy. For instance, WordNet [18], a full-scale lexical database, excludes polysemy from the description although it implements a host of other lexical and semantic relations. The presence of multiple word senses is quite typical rather than exceptional. In the Semcor corpus, for instance, Mihalcea and Moldovan [16] found 6.6 possible interpretations per word on average (using WordNet sense categorization). Even tiny sense variations are kept distinct in WordNet, and the database is probably as fine grained as possible. Mihalcea and Moldovan [16] point out that it is not uncommon that WN "word senses are so close together that a distinction is hard to be made even for humans" ([16]: 454).

2.2 Category Ambiguity

Category ambiguity is the most straightforward type of lexical ambiguity. This happens when a given word may be assigned to more than one grammatical or syntactic category as per context. One can find a number of such examples in Malayalam. For example, the word *pacca* "green" can be both noun and adverb. Similarly, the word *vEgam* can be both adverb and adverb. *kuTi* could be both verb and noun.

11. (a). *A sAriyuTe niRAm paccay-ANə* "The color of sari is green."
11. (b). *A pacca sAri vila-kkUTutal-ANə* "That green sari is costly."
12. (a). *avan vEgam vannu* "He came quickly."
12. (b). *A kArinRe vEgata kUTutal-ANə* "The speed of the car is more."

The words like *mukaLil* and *tAze* could be adverbs and postpositions.

13. (a). *avan AnayuTe mukLil kayaRi* "He climbed over the elephant."
13. (b). *avan mukaLil kayaRi* "He climbed up."
14. (a). *avan marattinRe tAze iRangi* "He climbed down from the tree."
14. (b). *avan tAze iRangi* "He climbed down."

Category ambiguities can be often be resolved by morphological inflection. For example, *kuTi* in *avan kuTikkunnu* "he drinks" is a verb and *kuTi* in *avan kuTi niRutti* "He stopped drinking" is noun. Frequently, ambiguity can be resolved by syntactic parsing. However, the problem increases when several categorical ambiguous words occur in the same sentence, each requiring being resolved syntactically.

2.2.1 Categorical Ambiguity Due to Historical Functional Reorganization

The inflected forms of nouns or verbs will denote different word category or functional category due to historical meaning change. For example, many of the postpositions in Malayalam are historically the inflected forms of verbs and nouns. The inflected forms *ninnu* "from," *paRRi* "about," *kuRiccu* "about," *koNTu* "by (means of)," *vaccu* "by (means of)," *cuRRi* "around," *nOkki* "towards," and *kUTe* "along with" are the inflected forms of the verb *nilkkuka* "stand," *paRRuka* "apply," *kuRikkuka* "aim," *koLLuka* "have," *vaykkuka* "to keep," *cuRRuka* "go around," *nOkkuka* "look at," and *kUTuka* "assemble," respectively. The postpositions *munne* "before" and *pinne* "after," *munpil* "in fornt," *pinnil* "at the back," *tAze* "below," and *mukaLil* "above" are the inflected or modified forms of nouns.

15. (a). *avan vITTil ninnum vannu* "He came from home."
15. (b). *avan vITTil nilkunnu* "He is (standing) the house."
16. (a). *avan enne paRRi paRanjnu* "He talked about me."
16. (b). *avan avaLuTe manasil kayaRi paRRi* "He landed on her heart."
17. (a). *avan enne kuRiccu paRanjnju* "He talked about me."
17. (b). *avan kaTalAsil kuRiccu* "He noted in the paper."
18. (a). *avan ate katti koNTu veTTi* "He cut it by a knife."
18. (b). *avanRe kaiyil veTTu koNTu* "He wounded his hand."
20. (a). *avan katti vaccu atine veTTi* "He cut it with a knife."
20. (b). *avan katti vaccirikkunnu* "He is having a knife."
21. (a). *vITTine cuRRi marngngaLANu* "There are trees surrounding the house."
21. (b). *avan vITTine cuRRi* "He went around the house."
22. (a). *avan avaLe nOkki vannu* "He came towards her."
22. (b). *avan avaLe nOkki* "He looked at her."
23. (a). *avan avaLuTe kUTe vannu* "He came with her."
23. (b). *avarellAm aviTe kUTi* "They gathered there."

Although people are sometimes said to be ambiguous in how they use language, ambiguity is, strictly speaking, a property of linguistic expressions. A word, phrase, or sentence is ambiguous if it has more than one meaning. Obviously, this definition does not say what meanings are or what it is for an expression to have one (or more than one). For a particular language, this information is provided by a grammar, which systematically pairs forms with meanings, ambiguous forms with more than one meaning.

3   Polysemy: Its Nature and Consequences

Here, we will be discussing the nature of polysemy, consequence of polysemy, and the ambiguity due to polysemy.

3.1   Polysemy

Polysemy (/pəˈlɪsɪmi/ or /ˈpɒlɪsiːmi/ from Greek: πολυ-, poly-, "many" and σῆμα, sêma, "sign") is the capacity for a sign (such as, a word, phrase, or symbol) to have multiple meanings (i.e., multiple semes or sememes and thus multiple senses), usually related by contiguity of meaning within a semantic field. Polysemy is thus distinct from homonymy – or homophony – which is an accidental similarity between two words (such as *bear* the animal, and the verb to *bear*); while homonymy is often a mere linguistic coincidence, polysemy is not.

Charles Fillmore and Beryl Atkins' [8] definition stipulates three elements:

(i)   The various senses of a polysemous word have a central origin.
(ii) The links between these senses form a network.
(iii) Understanding the "inner" one contributes to understanding of the "outer" one.

Polysemy is a pivotal concept within disciplines such as media studies and linguistics. The analysis of polysemy, synonymy, and hyponymy and meronymy is vital to taxonomy and ontology in the information-science senses of those terms. It has applications in pedagogy and machine learning, because they rely on word-sense disambiguation and schemas.

A polyseme is a word or phrase with different, but related senses. Since the test for polysemy is the vague concept of relatedness, judgments of polysemy can be difficult to make. Because applying preexisting words to new situations is a natural process of language change, looking at words' etymology is helpful in determining polysemy but not the only solution; as words become lost in etymology, what once was a useful distinction of meaning may no longer be so. Some apparently unrelated words share a common historical origin; however, so etymology is not a dependable test for polysemy, and dictionary writers also often defer to speakers' intuitions to judge polysemy in cases where it contradicts etymology. Malayalam has many polysemous words. For example, the verb *naTa* "walk" can mean "function," "occur," "happen," etc.; *OTu* can mean "run," "function," "occur," "flow," etc. [19].

In vertical polysemy, a word refers to a member of a subcategory (e.g., *kOzi* "fowl" is for denoting "chicken"). A closely related idea is metonymy, in which a word with one original meaning is used to refer to something else connected to it. Metaphorical uses of words are instances of polysemy as well. For example, *kuTi* "drink" might be specified as "drink as animate being." This can be metaphorically extended to inanimate beings as given in the following example:

24. *I vAhanam orupATu indhanam kuTikkunnnu*

  "This vehicle is consuming lot of petrol."

The difference between homonyms and polysemes is subtle. Lexicographers define polysemes within a single dictionary lemma, numbering different meanings, while homonyms are treated in separate lemmata. Semantic shift can separate a polysemous word into separate homonyms. For example, *naTa* as in *avan naTakkunnu* "he is walking" in which it originally denotes walking by means of legs. Now, it is used to denote functioning, occurring, and happening. The Malayalam lexicographer gives separate entries for these two senses. This is an example which shows how polysemy becomes homonymy. It can be inferred by psycholinguistic experiments that homonyms and polysemes are denoted differently in the mental lexicon of the people. It is often felt that the different senses of homonymous words interfere in the comprehension, whereas the same is not true with polysemes of a word. This argument has mixed results. One group of polysemes are those in which a word meaning an activity, perhaps derived from a verb, acquires the meanings of those engaged in the activity, or perhaps the results of the activity, or the time or place in which the activity occurs or has occurred. Sometimes, only one of those meanings is intended, depending on context, and sometimes multiple meanings are intended at the same time. Other types are derivations from one of the other meanings that lead to a verb or activity.

OTu "run"

(a) *OTu* – move fast using legs
(b) *OTu* – work as machine
(c) *OTu* – move as time

This example shows the specific polysemy where the same word is used at different levels of taxonomy. Example *a* contains *b*, and *b* contains *c*.

muTTə in Malayalam shows polysemy and homonymy. *muTTə* as a verb can mean "knock," "dash," "pressure for urinating, defecating, breathing trouble, finding difficulty in answering, etc. This is an instance for polysemy. *muTTə* as a known means "knee" and "support." This is an instance for homonymy.

The verbs in Malayalam are highly polysemous as only a handful amount of verbs are used to denoted innumerable number of events, actions, and processes. The verbs such as *iTuka, aTikkuka,* and *OTuka* are highly polysemous showing extensions of meaning in various dimensions from the core meaning.

A lexical conception of polysemy was developed by Atkins [3] in the form of lexical implication rules. These are rules that describe how words, in one lexical context, can then be used, in a different form, in a related context. A crude example of such a rule is the pastoral idea of "verbizing one's nouns." For example, certain nouns, used in certain contexts, can be converted into a verb, conveying a related meaning (e.g., In English, the nouns *chair, bench, table,* etc. are used in verbal sense.) This kind of verbalization of nouns into verbs is not found or very rare in Malayalam. In Malayalam, certain nouns such as *praNayam* "love (noun)" become verb by some formal change, that is, *prNayikkuka* "to love (verb)." A number of nouns borrowed from Sanskrit get converted into verbs in this fashion (*lAbham* "probit" > *lAbhikkuka* "make profit").

The polysemous nature of verbs helps Malayalam to form new verbs from the combination of noun and verb. In that case, the verbs act as verbalizers.

For example, *accu* "print" + *aTi* "beat" > *accaTi* "print"
kOppi-aTi (*kOppi* "copy" + *aTi* "beat" > *kOppiyaTi* "copy"

On a scale of meaning variance, ambiguity and vagueness are the two extremes, whereas polysemy is in between the other two. It shares features with both and is a common phenomenon in everyday language use. Polysemy involves lexemes that are clearly united (share a common schema) as well as clearly separable at the same time. Polysemous words are the result of lexemes gaining new usages over time which share the same phonological form and appear to have separate meanings to nonetymologists. *aTi* "beat" is one example of polysemy.

accaTi "print," *kaNNaTi* "wink," *kayyaTi* "clap," *muRRamaTi* "clean with broom," *ANiyaTi* "drive in a nail," *bhAgyamaTi*, "win a lottery"

Polysemy is sometimes mixed up with homonymy. However, there is a clear-cut distinction between these two. Polysemous lexemes always share the same etymological background and/or are conceived of as being semantically related by speakers, whereas homonymous words just happen to end up with the same phonological form. Therefore, homonymy may be seen as a subcategory of lexical ambiguity.

3.2   Test for Polysemy

Establishing polysemy is a challenging task. There are some tests for determining the presence of polysemy. In addition, polysemy is differentiated from other phenomena that involve potential multiplicity of meaning. A few potential cases of polysemy are to be explored. We have to deal with the polysemy paradox and consider ways in which types of polysemy can be characterized and categorized.

First, etymology is not a perfectly reliable arbiter of polysemy. A meaning that is etymologically related to another need not be appropriately similar to the initial meaning, as the two may drift apart over time to the point that they are no longer suitably related. An example of the verb *naTa* which originally denotes "moving by legs" is extended to denote "function" as in the following example:

25. *avanRe kaccavaTam nannAyi naTakkunnu*

"His business is going on well."

As seen from the above example, a word may acquire a new sense historically.

Second, it is arguable whether or not polysemy should be seen as coming in degrees. A bad argument for this conclusion goes as follows: polysemous terms enjoy related, similar meanings. Similarity comes in degrees. Therefore, polysemy comes in degrees. But the grade ability of a phrase does not follow from the grade ability of a condition for the correct application of the phrase. But whether or not polysemy comes in degrees, there are definitely cases in which it is vague whether or not two meanings for a term or phrase constitute polysemy. It will be often hard to tell whether or not a term is merely ambiguous or polysemous. The examples of *aTi, Otu,* and *naTa* will illustrate the above statement.

Zeugma is one of tests for polysemy. If one word seems to exhibit zeugma when applied in different contexts, it is likely that the contexts bring out different polysemes of the same word. If the two senses of the same word do not seem to fit, yet seem related, then it is likely that they are polysemous. The fact that this test again depends on speakers' judgments about relatedness, however, means that this test for polysemy is not dependable, but is rather merely a helpful conceptual aid. The following example will illustrate this statement.

26. *avaL avanuvENTi vITinREyum hrudayattinREyum vAtil tuRanniTTu*

"She waited leaving open her house door and heart"

Linguists and philosophers have developed various tests for polysemy and ambiguity. Most of these tests involve attempts to "freeze" a single meaning of the putatively polysemous phrase and then see if that frozen meaning can be used to express the multiple meanings. Unfortunately, this makes most tests unsuitable for distinguishing polysemous phrases from merely ambiguous phrases, since both enjoy multiple meanings.

These are tests, not knockdown arguments. Most of them require a somewhat slippery notion of interpretation. One may wonder whether the sentence in question expresses multiple meanings or not. Of course, with some creativity and perhaps some practice, competent language users can find themselves detecting meanings in unexpected places. What the tests presuppose is that the reader can access something like interpretations of a sentence that are not deviant.

4   Resolving Polysemy

This is the main focus of the chapter. Here, we will be discussing about the resolving polysemy by contextual similarity.

4.1   Relevance of Context for Resolving Polysemy

According to Rumshisky ([24]: 217), the combination of the following two factors assigns meaning to a word: the syntactic frame of a word and the semantics of the word in that syntactic-dependent frame. Such words are referred as "selector" by Rumshisky ([24]: 217). This is applicable for both the head word and dependent word. The syntactic frame is extendable to the minor categories such as adverbials, locatives, and temporal adjuncts and to the subphrasal cues such as genitives, partitives, negatives, bare plural/determiner distinction, and infinitives ([24]: 217). The set of all "usage contexts" in which a polysemous word occurs can usually be divided into groups. Each group roughly corresponds to a distinct "sense."

Consider following sentences with the verbs *niSEdhikkuka* "deny" made use of in 27a and *paRayuka* "say" made use of in 28a to illustrate the contribution of different context parameters to disambiguation. The difference in the syntactic patterns for the verb *niSEdhiccu* "deny" as shown in 27b and 27c disambiguate between the two dominant senses: "refuse to grant" and "proclaim false" and similarly the difference in the syntactic patterns of *paRayuka* "say" as shown in 28a and 28b disambiguate between "complain" and "blame."

Syntactic frame

27. (a). *adhikAri atinu kAraNam uNTenna kAriyatte nishEdhiccu {enna-clause)*

"The authority denied that there is a reason for that" (proclaim false)

27. (b). *adhikAri kAraNatte nishEdhiccu* [NP]

"The authority denied the reason." (proclaim false)

27. (c). *adhikAri rAdhaykku visa nishEdhiccu* [NP]

"The authority dined visa to Radha." (refuse to grant)

28. (a). *avaL koccine aTiccu ennU avan kuRRam paRanjnu* [ennu-clause]

"He complained that she bet the child." (complain)

28. (b). *avan avaLe kuRRam paRanjnju* (blame) [NP]

"He blamed her."

Consider the following sentences with *OTuka* "run," *valicceTukkuka* "absorb," and *perumARuka* "behave." The contrasting argument and/or adjunct semantics shown in *29, 30,* and *31* evoke the different the senses of *OTuka* "run," *valicce-Tukkuka* "absorb," and *perumARuka* "behave," respectively. The relevant argument type is shown in brackets and the corresponding sense is shown in parentheses. The following examples show the semantics of the arguments and adjuncts/adverbials:

29. (a). *pOlIs kaLLanRe puRakil OTi*

"The police ran after the thief." [chased person] (run)

29. (b). *avaL avanRe kUTe OTippOyi*

"She ran away with him." [accompanying person] (elope)

30. (a). *panjnji veLLatte valicceTuttu*

"The cotton absorbed the water." [water] (absorb}

30. (b). *avaL avnRe kaiyil ninnum paisa valicceTuttu* [money] (take away)

"She extracted money from him."

31. (a). *avan avaLOTU mariyadayAyi perumARi*

"He behaved with her with respect." [with respect] (behave)

31. (b). *pOlIs avane nallavaNNam perumARi*

"Police beat him severely." [severely] (beat)

A lexicographer establishes a set of senses available to a particular lexical item and (to some extent) specifies the context elements which typically activate each sense. This procedure is formalized in several current resource-oriented projects such as FrameNet and Corpus Pattern Analysis (CPA). The semantics of the arguments is represented in FrameNet by making use of Fillmore's case roles. Hanks and Pustejovsky [12] attempt to arrange prototypical norms of usage for individual words in terms of contextual patterns by CPA.

4.2 Polysemy and Distributional Similarity

A number of tasks in NLP make use of the notion of distributional similarity. Distributional similarity is exploited in the areas such as word-sense disambiguation (WSD), sense induction, automatic thesaurus construction, selectional preference acquisition, and semantic role labeling ([24]: 220). Semantically similar words (as in thesaurus construction) or similar uses of the same word (as in WSD) and sense induction are identified by making use of distributional similarity. Distributional similarity results in clusters of distributionally similar words. The problem of data sparsity faced by many NLP tasks is addressed by these clusters. The distributional similarity of a polysemous word gives rise to the generalization that must be applied to different "senses" rather than to its entire occurrence regularly. The external knowledge sources such as FrameNet, machine-readable dictionaries, and WordNet give rise to information that represents semantics of the arguments ([24]: 220).

A feature vector represents a context. Each feature vector is mapped to some context component. The target word along with the context component gives the frequency of occurrence of the value of each feature vector. All the context structures or the probability of co-occurrence distribution represent the feature vector. Different approaches make use of different methods. Widdows and Dorrow [28] and Agirre et al. [1] make use of co-occurrence graph. Schütze [26], Gale et al. [10], and Widdows and Dorrow [28] make use of "distributional features based on bag-of words style co-occurrence statistics." Grefenette [11], Lin [14], and Pantel and Lin [21] make use of context structures incorporating syntactic information and sometimes semantic information from external sources.

From the distributional exemplification of the target word, the occurrences corresponding to each sense can be separated out. This will lead to the resolving of polysemy. According to Schütze [26], Grefenette [11], and Lin [14], clustering similar occurrence context for each word or clustering of actual words which are distributionally similar resolves this problem.

4.3 Sense Assignment for Polysemous Verbs

Sense inventories for polysemous verbs are often comprised by a number of related senses. Computational approaches to word-sense disambiguation assign a sense to each word in an utterance from an inventory of senses. This simplified statement may not be true when the meaning of a complex utterance is computed. Take, for example, a target verb showing polysemy with certain semantic inclinations. Different sematic components select different senses of a verb in a given argument position. In the sentence 29b, the "elope" sense of *OTuka* is selected by the argument indicating the accompanying person while "run," sense in 6a is selected by the argument indicating the "chased person"; similarly, in 30b, the "take away" sense is selected by the argument denoting "money" in the direct object position, whereas the "absorb" sense in 30a is selected by the argument denoting the direct object "water." In the same way in 31b, the "beat" sense is selected by the adverbial adjunct "*nallavaNNam*" "severely" and the "behave" sense in 31a is selected by the adverbial adjunct *maryAdayAyi* "with respect."

4.3.1 Sense-Activating Argument Sets for a Polysemous Verb

The same meaning of a verb can be triggered by a number of semantically different arguments. Certain pertinent semantic feature will be central to the interpretation of the meaning for some of the verb. Other verbs are allowed merely by a contextual interpretation. An ad hoc semantic category will be prompted by each sense of the target verb in the relevant argument position. For example, consider the senses of the verb *eTukkuka* "take (something)," "raise (hood like a snake), "take photo," "score," "copy," etc. The following examples exemplify the lexical items that occur in the position of direct object.

32. (a). *pAmpU paTam eTuttu* "The snake raised its hood."
32. (b). *avan avaLuTe paTam eTuttu* "He photographed her."
32. (c). *avan atinRe pakarppU eTuttu* "He took the copy of it."
32. (d). *avan kaNakkinU nURu SatamAnam mArkkU eTuttu*

"He scored hundred percent mark in mathematics."

The verbs need to be activated or evoked for a relevant sense by its arguments. The semantically quite discrete nouns in each argument set activate the relevant sense of the verb. A particular aspect of the sense is selected by the context provided by the verb. Mostly, an argument set carry a central component of their meaning as well as certain other peripheral components. The argument set has core members which are polysemous. In order to activate the appropriate sense of the verb, a "bidirectional selection" process needs to be implemented. But notice that the interpretation of 33a and 33b, for example, is quite different.

33. (a). *pOlIs avane kasRRaTiyil eTuttu* "The police arrested him."
33. (b). *avan A jOli ceyyAn kuRaccu samayam eTuttu*

"He took some time to do the work."

In the above sentences, both the words in the argument position activate the same sense of *eTukkuka*. But the disambiguation is between the EVENT reading and the TIME reading. It has to be noted that the diverse dependencies the word enters into rely on the different aspects of the meaning. For example, consider the use of the noun arguments with the verb *eTukkuka* in the sentences *34a* and *34b* given below.

34. (a). *kamsanRe kArAgrhattil krishNan janmam eTuttu*

"Krishnan was born in Kamsan's jail."

34. (b). *innale kOTatiyil ninnum avane jAmyattil eTuttu*

"He has been bailed out from the court."

In the above examples, the words *janmam* "birth" and *jAmyattil* "bail-in" activate different senses for *eTukkuka* "take." The context provided by the verb effectively changes the relevant semantic components in the interpretation of the senses.

4.3.2   Sense Separation Based on Selector

In the case of homonymy, sufficiently distinctive semantic features are chosen for denoting different senses ([24]: 224). The relevant lexical items can be grouped together by making use of the overall distributional similarity between arguments. For example, movement sense of *maTangnguka* "return" is easily differentiated from the cluster of senses associated to "fold as paper."

35. (a). *avaL OphIsil ninnum vITTilEkkU maTangi*

"She returned from office."

35. (b). *pustakattinRe pEjukaLokke maTngippOyi*

"The pages of the book got folded up."

Similarly, the movement sense of *naTakkuka* "walk" is easily differentiated from the cluster of senses associated to "go on."

36. (a). *kunjnju patukkeppatukke naTakkAn tuTangi*

"The child started walking slowly."

36. (b). *aviTatte tiyETTaRil nalla cinima naTakkunnu*

"A cinema is going on in the theatre."

In the case of polysemy, differentiating dissimilar meanings of the verb is extremely difficult. This problem has been the subject of extensive study in lexical semantics. It aims at addressing the question of selecting distinct senses based on context. There is no clear-cut method to say when a context selects a distinct sense or when a context simply modulates the meaning [2, 7, 23]. This is crucial for the computational method of word-sense disambiguation. Lexicographers often face problems in determining when to define a set of usages as a distinct sense while "lumping and splitting" senses during dictionary construction. The sense separation is frequently determined on ad hoc basis. It results in instances where the same occurrence may fall under more than one sense category simultaneously resulting in numerous cases of "overlapping senses." Resolving verbal polysemy often runs into this problem of indecision.

There is a necessity to decide what sense is activated by which selector. While resorting to such attempt, we should have in mind that at least some of the meanings of the verb are interconnected. The incidences of a polysemous verb in a corpus cluster into certain number of groups, each roughly corresponding to a sense ([24]: 225). There are a lot of instances in which the meaning dissimilarities are straightforward and easily noticeable. But there are some boundary cases where the sense diction of the verb is not clear cut. According to Rumshisky ([24]: 225), three kinds of selectors are possible in a given argument position: (i) Good disambiguators: Here, one meaning of the target word is instantaneously selected by the selector. (ii) Poor disambiguators: Here, the selectors may choose either sense; more contexts may be needed to disambiguate themselves. (iii) Boundary cases: Here, it is impossible to make the choice between two senses of the target word. For instance, for the verb *kANikkuka*, "show" in *37*, *sarve* "survey," and *paTam* "photo" are good disambiguators, whereas *grAph* "graph" is a clear example of a boundary case.

37. (a). *A paTam avaLuTe mughatte nalla vaNNam kANikkunnu* ("picto rially represent")

"That picture shows her face well."

37. (b). *A sarve vyvasAya mEkhalayil uLLa sarkkArinRe purOgatiye kANikkunnu*

"The survey shows the improvement of government in industrial sector." ("demonstrated by evidence or argument")

37. (c). *I grAph I mAsattil uLLa maZyuTe SarASari aLavine kANikiunnu*

"The graph shows an average rainfall in this month." (both senses?)

Each individual sense needs to be clearly defined for the identification of boundary cases. Such instances are better interpreted as examples of "multiple selection" (i.e., simultaneous initiation of both meanings); they are not simply the examples of definitions of overlapping meaning [25]. Even syntactic structure cannot always supersede the elucidation inherent to some selectors. For instance, in *38*, it is almost impossible to resolve *nishEdhikkuka* "deny" between "refuse to grant" and "proclaim false."

38. (a). *vayOdhikarkkU avaruTe stAnam nishEdhikkappeTunnu*

"Elders are often denied the status of adulthood."

38. (b). *cila jAdikkAr strIkaLkkU svAtandryam nishEdikkunnu*

"People of certain caste denies autonomy to women."

Instead, in *39*, the selector itself is polysemous, with two elucidations obtainable for it, and there is a requirement for disambiguation by context before it can trigger the suitable meaning of the verb.

39. (a). *paNTatte aphiprAytte nishEdikku*

"Deny the traditional view." ("proclaim false")

39. (b). *avanRe anuvAdam nishEdikku*

"Deny him permission." ("refuse to grant")

The succeeding sections elaborate on the scheming of a computational approach for automatic meaning detection.

4.4   Contextual Similarity and Dissimilarity

By means of contextual similarity and dissimilarity, we can assign senses for a verb automatically text. Based on the specific task, the decision that one context is similar to the other may vary. If our aim is to find out the meaning of a verb based on its arguments, it can be inferred that different contextual arguments give different senses for the concerned verb.

For instance, the verb *eTukkuka* "take" and the direct object relation define a particular context of incidence for the noun. At its most basic, contextual similarity between occurrences of the concerned verb should reflect to what extent the contexts in which the verb occur will overlap. Similarity between contexts may be conveyed by the similarity of their arguments.

40. (a). *avan kuTTiye kaiyil eTuttu* "He took the child in his hand."
40. (b). *avan kalline kaiyl eTuttu* "He took the stone in hand."
41. *avan kaNakkinu nURu mArkku eTuttu*

"He scored hundred marks."

It can be inferred from the sentences *40a* and *40b* that the word *eTu* occurring in the two instances overlap in the context of their occurrence, thereby inferring the same sense. In sentence *41, eTu* occurs in a dissimilar context, thereby inferring a different sense.

4.5   Algorithm Architecture

The similarity and dissimilarity between contexts is made use of here to separate the senses of a verb. But, computing similarity and dissimilarity between contexts poses a problem. Two or more contexts must be similar with respect to their selectional properties, that is, the same semantic components in the specified argument position. Similarity and dissimilarity between contexts are achieved in the present work of WSD by neural network approach.

4.6 Neural Network Approach

Automated language understanding requires the determination of the concept which a given use of a word represents. This process is referred to as word-sense disambiguation (WSD). WSD is typically affected in natural language processing systems by utilizing semantic feature lists for each word in the system's lexicon, together with restriction mechanisms such as case role selection. However, it is often impractical to manually encode such information, especially for generalized text where the variety and meaning of words is potentially unrestricted. Furthermore, restriction mechanisms usually operate within a single sentence, and thus, the broader context cannot assist in the disambiguation process.

WSD is executed using neural network approaches suggested by Cottrell and Small [6] and Waltz and Pollack [27]. The nodes ("neurons") representing words or concepts connected by "activatory" links make networks which are found in these models: the words activate the concepts to which they are semantically related and vice versa. The "lateral" inhibitory links usually interconnect competing senses of a given word. The nodes corresponding to the words in the sentence to be analyzed are activated initially. The adjacent ones of these words are activated in the next cycle in turn; then, the immediate adjacent ones of these neighbors are stimulated, and so on. Using a parallel, analog, relaxation process, the network stabilizes in a state in which one sense for each input word is more activated than the others after a number of cycles.

Neural network approaches to WSD seem able to capture most of what cannot be handled by overlap strategies such as Lesk's [13]. However, the networks used in experiments so far are hand coded and thus necessarily very small (at most, a few dozen words and concepts). Due to a lack of real-size data, it is not clear that the same neural net models will scale up for realistic application. Further, some approaches rely on "context setting" nodes to prime particular word senses in order to force the correct interpretation. But as Waltz and Pollack point out, it is possible that such words are not explicitly present in the text under analysis, but may be inferred by the reader from the presence of other related words.

To solve this problem, words in such networks have been represented by sets of semantic "micro-features" [5, 27] which corresponds to fundamental semantic distinctions, characteristic duration of events, locations, and other similar distinctions that humans typically make about situations in the world. Each concept in the network is linked, via bidirectional activatory or inhibitory links, to only a subset of the complete micro-feature set. A given concept theoretically shares several micro-features with concepts to which it is closely related and will, therefore, activate the nodes corresponding to closely related concepts when it is activated itself.

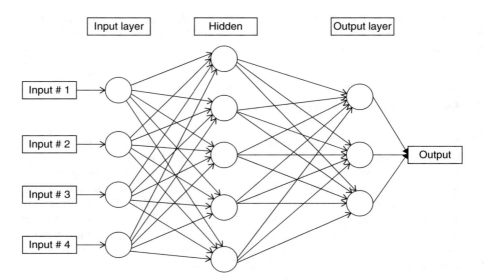

Fig. 1 Neural network disambiguation example

The words are complex units in the model we have adopted. Each word in the input is represented by a word node connected by excitatory links to sense nodes representing the different possibilities. Each sense node is in turn connected by excitatory links to word nodes representing the words in tile definition of that sense. This process is repeated a number of times, creating an increasingly complex and interconnected network. In the network, all words are reduced to their lemmas and grammatical words are excluded. The different sense nodes for a given word are interconnected by lateral inhibitory links (Fig. 1).

When the network is run, the input word nodes are activated first. Then each input word node sends activation to its sense nodes, which in turn send activation to the word nodes to which they are connected, and so on throughout the network for a number of cycles. At each cycle, word and sense nodes receive feedback from connected nodes. Competing sense nodes send inhibition to one another. Feedback and inhibition cooperate in a "winner-take-all" strategy to activate increasingly related word and sense nodes and deactivate the unrelated or weakly related nodes. Eventually, after a few dozen cycles, the network stabilizes in a configuration where only the sense nodes with the strongest relations to other nodes in the network are activated. Because of the "winner-take-all" strategy, at most one sense node per word will ultimately be activated.

4.7   Analysis of Annotation Decisions

In order to identify the senses of a verb, say, for example, *OTu* "run," a simple recurrent neural network-based learning approach is applied. It is from the human understanding of word from its previous word in the sequence recurrent neural network (RNN) takes its idea. For the understanding of more about the present events based on the previous events, earlier neural network did not have the mechanism. The loops of neural networks which are RNN allow the information to exist in it. A simple RNN model is shown in Fig. 2.

In Fig. 2, "NN" denotes any neural network architecture, "x_t" denotes an input, and "h_t" denotes the output. Output value to the next phase is passed by loop. The present information with the immediate previous value is captured essentially with the help of RNNs. However, if the immediate previous value or values do not contribute for the understanding of the sequence of words, it will be omitted. An extended version of RNN long short-term memory is used in such cases, which can even capture long dependencies. In order to distinguish the nine sense classes of *OTu*, a simple RNN is used.

The evaluation measures acquired for nine epochs are shown in Tables 2 and 3. The network can be modified to get better results by creating more words and its corresponding senses.

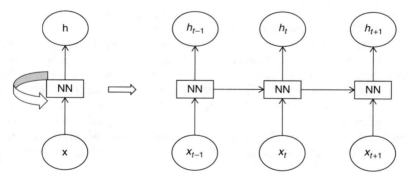

Fig. 2 Simple RNN

Table 2 No. of data in each class

Sense No.	No. of data
1	1001
2	1099
3	1045
4	1057
5	1100
6	1568
7	1154
8	1326
9	1049

Table 3 Result of nine epochs, each epoch is run for 1000 steps

Accuracy	Precision	Recall	F-measure
0.34328	0.368	0.343	0.332
0.69403	0.765	0.694	0.701
0.93284	0.938	0.933	0.933
0.97015	0.972	0.97	0.97
0.97761	0.981	0.978	0.978
0.85075	0.876	0.851	0.856
0.90299	0.905	0.903	0.902
0.98507	0.985	0.985	0.985
0.99254	0.993	0.993	0.993

5   Conclusion

The result of the proposed method is encouraging as shown Tables 1 and 2. Each of the target senses is associated with a cluster of selectional equivalents for that sense. The selectional equivalents are presented as contextualized vectors of dependable selectors. The clusters resulted from this are used to find out the selectors that trigger each sense. The association score obtained for each selector indicates which sense it tends to trigger. We are able to resolve some of the problems faced by the previous efforts to resolve polysemy. The evaluations discussed in the result section show that simple RNN provides a competing result for identifying the senses of *OTu* "run." This approach can be improved if we have a huge corpus covering all the senses of verbs. Such corpus is very difficult to get. Malayalam does not have a corpus which can be made use of for NLP works such as WSD. If such corpus is available, we can automatically classify the different senses of a verb by means of contextual clusters. We are able to identify the possible number of senses for a verb automatically.

References

1. Agirre, E., et al. (2006). *Two graph-based algorithms for state of the art WSD*. Proceedings of the 2006 Conference on Empirical Methods in Natural Language Processing 22–23 (July 2006), Sidney, Australia, pp. 585–593.
2. Apresjan, J. U. (1973). Regular polysemy. *Linguistics, 142*(5), 5–32.
3. Atkins, S. (1991). Tools for computer-aided corpus lexicography: The Hector project. *Acta Linguistica Hungarica, 41*, 5–72.
4. Böhmerová, A. D. A. (2010). Lexical ambiguity as a linguistic and lexicographical phenomenon in English, with comparisons to Slovak. In J. Kehoe (Ed.), *Ambiguity, Conference Proceedings* (pp. 27–35). Verbum, Ružomberok.
5. Bookman, L. A. (1987). *A microfeature based scheme for modeling semantics*. Proceeding of the 10th International Joint Conference on Artificial Intelligence, IJCAI'87, Milan, Italy, pp. 611–614.
6. Cottrell, G. W., & Small, S. L. (1983). A connectionist scheme for modelling word sense disambiguation. *Cognition and Brain Theory, 6*, 89–120.

7. Cruse, D. A. (1995). Polysemy and related phenomena from a cognitive linguistic viewpoint. In *Computational lexical semantics*. Cambridge: Cambridge University Press.
8. Fillmore, C. J., & Atkins, B. T. S. (1991). *Invited lecture presented at the 29th Annual Meeting of the Association for Computational Linguistics*, 18–21 June 1991, Berkeley, CA.
9. Firth, J. R. (1957). *Papers in linguistics 1934–1951*. Oxford: Oxford University Press.
10. Gale, W. A., Church, K. W., & Yarowsky, D. (1993). A method for disambiguating word senses in a large corpus. *Computers and the Humanities, 26*, 415–439.
11. Grefenette, G. (1994). *Exploration in automatic thesaurus discovery*. Norwell, MA: Kluwer Academic Publishers.
12. Hanks, P., & Pustejovsky, J. (2005). A pattern dictionary for natural language processing. *Revue Française de Linguistique Appliquee, 10*(2).
13. Lesk, M. (1986). Automatic sense disambiguation using machine readable dictionaries: How to tell a pine cone from an ice cream cone. In V. DeBuys (Ed.), *Proceedings of the 5th Annual international Conference on Systems Documentation (Toronto, ON, Canada)*. SIGDOC, June 1986, pp. 24–26.
14. Lin, D. (1998). *Automatic retrieval and clustering of similar words*. Proceedings of the 17th International Conference on Computational Linguistics, Montreal, Canada.
15. Lyons, J. (1977). *Semantics volumes I and II*. Cambridge: Cambridge University Press.
16. Mihalcea, R., & Moldovan, D. (2001). *EZ.WordNet: Principles for automatic generation of a coarse grained WordNet*. Proceedings of FLAIRS, Key West, FL, pp. 454–459.
17. Miller, G. A., & Charles, W. G. (1991). Contextual correlates of semantic similarity. *Language, Cognition and Neuroscience, 6*(1), 1–28.
18. Miller, G. A., Beckwith, R., Fellbaum, C. H., Gross, D., & Miller, K. J. (1990). Introduction to WordNet: An on-line lexical database. *International Journal of Lexicography, 3*(4), 235–244.
19. Mohan Raj, S. N., Sachin Kumar, S., & Rajendran, S. (2017). Resolving polysemy in Malayalam verbs. *Language in India* wwwlanguageinindia.com ISSN 1930-2940, 17:1, pp. 153–163.
20. Mohan Raj, S. N., Sachin Kumar, S., Rajendran, S., & Soman, S. K. P.. (2019). Word sense disambiguation of Malayalam nouns. *Recent advances in computational intelligence*. ISSN 1860-949X ISSN 1860-9503 (electronic) *Studies in computational intelligence*. ISBN 978-3-030-12499-1 ISBN 978-3-030-12500-4 (eBook). Springer Nature Switzerland AG 2019.
21. Pantel, P., & Lin, D. (2002). *Discovering word senses from text*. Proceedings of ACM SIGKDD Conference on Knowledge Discovery and Data Mining, pp. 613–619.
22. Pehar, D. (2001). Use of ambiguities in peace agreements. In J. Kurbalija & H. Slavik (Eds.), *Language and diplomacy* (pp. 163–200). Malta: DiploFoundation.
23. Pustejovsky, J. (1995). *The generative lexicon*. Cambridge, MA: MIT Press.
24. Rumshisky, A. (2008). *Verbal polysemy resolution through contextualized clustering of arguments* (Brandeis University Ph.D).
25. Rumshisky, A., Grinberg, V. A., & Pustejovsky, J. (2007). Detecting selectional behavior of complex types in text. In B. Pierrette, L. Danlos, & K. Kanzaki (Eds.), *Fourth international workshop on generative approaches to the lexicon*. Paris, France.
26. Schütze, H. (1998). Automatic word sense discrimination. *Computational Linguistics, 24*(1), 97–123.
27. Waltz, D. L., & Pollack, J. B. (1985). Massively parallel parsing: A strongly interactive model of natural language interpretation. *Cognitive Science, 9*, 51–74.
28. Widdows, D., & Dorrow, B. (2002). *A graph model for unsupervised lexical acquisition*. 19th International Conference on Computational Linguistics, August, 2002, Taipei, Taiwan, pp. 1093–1099.

A Clustering-Based Optimized Stable Election Protocol in Wireless Sensor Networks

Samayveer Singh

1 Introduction

Over the last few decades, there has been a significant evolution in academia as well as the industry in the field of wireless sensor networks (WSNs) due to tremendous advancement in the sensor nodes abilities like sensing, computation, and communication capabilities. The WSNs possess different characteristics like power consumption constraints for nodes, ability to survive with node failures, homogeneity and heterogeneity of nodes, scalability to the large scale of deployment, ability to withstand harsh environmental conditions, ease of use, and cross-layer design. In WSN, various sensors are spread in the target area for monitoring and recording the physical terms of the environment that have a small size, low processing power, less storage, and low energy abilities [1]. These sensor nodes can sense the target and make an infrastructure-less wireless communication among them and base station (BS). The sensors accumulate the data from the target area and forward it to the BS directly or with the assistance of other sensor nodes. The BS is connected to the server or other sensor networks with the help of wired/wireless links. Thus, the information is further sent to the end user via the internet. The BS is supposed to be reliable and is also capable of performing any operation. The WSNs are being used in different applications like battlefield surveillance, healthcare monitoring, forest fire detection, landslide detection, water quality monitoring, natural disaster prevention, structural health monitoring, data center monitoring and data logging, environmental conditions such as temperature, sound, pollution levels, humidity, and wind, industrial and consumer applications like machine health monitoring, industrial process monitoring and control, and so on [2].

S. Singh (✉)
Department of Computer Science & Engineering, Dr B R Ambedkar National Institute of
Technology, Jalandhar, Punjab, India

© Springer Nature Switzerland AG 2021
R. Kumar, S. Paiva (eds.), *Applications in Ubiquitous Computing*, EAI/Springer
Innovations in Communication and Computing,
https://doi.org/10.1007/978-3-030-35280-6_8

On the basis of the above-discussed applications, we can categorize the wireless sensor node deployment into two groups like deterministic and nondeterministic. In deterministic deployment, sensor nodes are placed in a controlled manner or manually at the selected locations where the deployment area is physically accessible such as city sense monitoring, soil monitoring, nursery monitoring, grid deployment, etc. On the other hand, in nondeterministic deployments, sensor nodes are deployed into physically inaccessible areas using other sources like sensors that are dropped from an aircraft, for example, battlefield surveillance, different harsh environment where a human cannot be reached, and landslide detection, etc. The nondeterministic deployment is also called as random deployment. In that, sensor nodes are deployed through some aircraft. Now, we discuss the very first protocol called low-energy adaptive clustering hierarchy (LEACH). It is also called hierarchical routing protocol [3] and it is used in the theory of data accumulation in which network is divided into different groups or clustered. Data aggregation is used for enhanced performance in terms of a lifespan of the system. There is a substantial similarity between the communication architecture of the cellular network and LEACH. It is established on the signal amplitude, and a router works as a cell header node or cluster head. There is a specific cluster head (CH) election method that is based on a probability function, and it has many criteria that are based on the preferred percentage of CH and the many times for the duration of which a node acquired the role of CH. In this algorithm, there is a predetermined time interval which is termed as a round where every round has two phases: the first one is an initialization phase and the second one is a transmission phase. Nowadays, the research community is following the clustering of sensor nodes to accomplish the aim of scalability of the network. The clustering techniques can be divided into two parts: hierarchical clustering and clustering partitioned. The method of clustering partitioned is more well organized and faster than the hierarchical classification and consortium is partitioned based on solid suppositions in the study cited herein [2]. In the perspective of clustering of nodes mostly used partition techniques are the clustering using fuzzy logic, k-means, c-means algorithm, etc.

In this work, we propose an optimized stable election protocol for enhancing the lifespan of a network by considering a fuzzy-based clustering technique along with the chain-based data collection and data aggregation process. It considers three parameters for effective clustering, namely, residual energy of a sensor node, node density within a cluster, and the distance between the sensor node and the base station. The proposed method considers a chain-based data gathering and transmission process for intra- and intercluster communication. A data aggregation process is also introduced for removing the redundant data which helps in decreasing the transmission cost and overhead of the networks. This work considers the homogeneous and heterogeneous networks for evaluating the performance of the proposed protocols.

The rest of the work is organized as follows: Literature review is discussed in Sect. 2. Section 3 discusses the network, energy model, and radio energy dissipation model. Section 4 deliberates the fuzzy logic system for the proposed protocol, and a chain-based data collection and data aggregation processes are discussed in Sect. 5. Section 6 deliberates the simulation results and their discussions, and finally, the paper is concluded in Sect. 7.

2 Literature Review

One of the major issues in WSNs is limited power supply, the limited size of sensor nodes, no rechargeable, etc., because these are battery operated. We need to design an energy-efficient protocol for solving such type of problems. A lot of clustering techniques have been proposed for load dissemination between the sensors in the past two decades. The load dissemination among the sensors provides a solution for the energy constraints issue of the WSNs and to extend the network lifetime. Low-energy adaptive clustering hierarchy (LEACH) is the very first distributed clustering protocol which has two-phase implementations, namely, setup phase and steady-state phase [3]. In the first phase, nodes are divided into clusters and their respective cluster heads are selected, and in the second phase, data are transmitted. In the LEACH, the cluster heads are selected randomly based on a probability function and every node gets a chance to become the CH to balance the energy consumption. Power-efficient gathering in sensor information systems (PEGASIS) is an extension of LEACH which forms chains using the sensors in the networks [4]. In the chains, the farthest node starts to collect the data and sends it to its nearer node, and then, this nearer node sends data to its next nearer node. This process continues until BS receives the data. This protocol is not appropriate for enormous networks because global information requires the available nodes in the networks.

The heterogeneous stable election protocol (hetSEP) is discussed for two and three-level heterogeneity [5] to enrich the lifetime of the network. It helps in the election of the CHs and their corresponding cluster members by weighted election probability and threshold function. It increases the overhead near the sink when the data transmission distance is too long. The distributed energy-efficient clustering (DEEC) protocol considers two-level and multilevel energy heterogeneous networks in WSNs [6]. The DEEC selects cluster heads using the ratio of residual energy of each node and the average energy of the network. It does not use extra energy of higher level nodes efficiently because the energy of the nodes is randomly allocated from a given energy interval. Thus, it may not be feasible to design such a network. In [7], three-level heterogeneity models are discussed in which the CHs are decided based on the residual energy of sensors to enrich the lifetime of the network. It helps in cluster heads' and cluster members' election by its weighted election probability and threshold function. However, it needs extra energy to rebuild the clusters in any iteration. In this protocol, data may be lost if cluster heads are not able to communicate with each other. Maheswari et al. discuss node degree-based energy efficient two-level clustering for wireless sensor networks to minimize the intercluster energy [8]. It considers the selection of normal cluster head by using node degree, centrality, and battery power. This method employs multi-hop communication and consequently suffers from the unexpected load at the sink node. The papers [9–11] discuss cluster techniques for the heterogeneous wireless sensor networks. They have considered the various levels of heterogeneity in WSNs. These papers do not consider the chaining approach for data collection and do not consider the data aggregation process between the sensor nodes and the cluster head nodes. Singh et al. discuss a protocol HEED-FL protocol which considers the basic approach of the HEED [12]. It selects the cluster head by using fuzzy logic based on the residual energy parameter of the sensor nodes and distance between a sensor and base station parameter. This paper suffers from the load balancing in the cluster head selection process and data aggregation process at the time of data collection.

Faisal et al. discuss a zonal-stable election protocol for WSNs for hybrid routing called Z-SEP [13]. It uses the same cluster formation as discussed in LEACH [3]. It does not consider normal nodes in the clustering process. The Z-SEP [13] is extended by Khan et al. as AZ-SEP [14] in which the authors discuss a hybrid and multi-hop advanced zonal-stable election protocol for WSN that communication of sensors with the BS is hybrid [14]. In this method, certain sensors communicate directly, while others use clustering mechanism to transmit data. The complete monitoring field is divided into three zones based on their nodes' energy and introduced a new mechanism of cluster head selection based on residual energy and distance from the BS. This method suffers a load-balancing problem within the cluster sets. Smaragdakis et al. discuss a heterogeneous protocol to prolong the time interval before the death of the first node called SEP [15]. In SEP, the election of cluster heads is based on weighted election probabilities of each node and considers the residual energy of the individual nodes. SEP considers two types of sensor nodes to define the heterogeneity of the networks.

In paper [16], a network model has been proposed which incorporates heterogeneity in terms of their energy. This model contains three-tier node heterogeneity, namely, tier-1, tier-2, and tier-3 heterogeneity. In paper [17], a 3-level heterogeneous network model for WSNs to enhance the network lifetime, which is characterized by a single parameter, is discussed. Depending upon the value of the model parameter, it can describe 1-level, 2-level, and 3-level heterogeneity. This heterogeneous network model also helps to select cluster heads and their respective cluster members by using weighted election probability and threshold function. The performance of the proposed model is computed in terms of the network lifetime by implementing DEEC protocol in [17]. In the papers cited herein [18, 19], a deterministic energy-efficient protocol for an adjustable sensing range is considered the underlying network of heterogeneous nature. The heterogeneous network model is parameterized that has 3-level heterogeneity. The proposed model can describe 1-level, 2-level, and 3-level heterogeneity. As the level of heterogeneity increases, the network lifetime increases. Furthermore, decreasing the value of the model parameter increases the network lifetime. Manju et al. [20] propose an energy-efficient scheduling algorithm based on learning automata for target coverage problem. The learning automata-based technique helps a sensor node to select its appropriate state (either active or sleep). To prove the effectiveness of their proposed scheduling method, they conduct a detailed set of simulations and compare the performance of their algorithm with the existing algorithms. Manju et al. [21] propose a new energy-efficient heuristic to schedule the sensors in different non-disjoint sensor covers which helps to maximize network lifetime. At first, the authors' heuristic identifies all the critical targets (least covered) and the critical sensors (covering critical targets). The critical targets, covered by the minimum number of sensors, will be the targets that become uncovered first. Utilizing critical sensors efficiently will help to increase the network lifetime. In their method, they try to select the minimum number of critical sensors in each sensor cover so that the critical targets can be covered for a longer period. Mehra et al. discuss an enhanced clustering algorithm based on fuzzy logic E-CAFL which is an improvement over the CAFL protocol [22]. The E-CAFL takes account of the residual energy, node density in its locality, and distance from the sink and feeds into the fuzzy inference system. A rank of each node is computed for the candidature of cluster coordinator. The results illustrate better performance instability period and protracted lifetime. In the next section, we will discuss the fuzzy logic system which will help in efficient clustering.

3   Network, Energy, and Radio Dissipation Model

In this section, we consider the following assumptions for the design of wireless sensor networks.

- All the sensors are stationary, and they have their own unique identity with GPS enable technology.
- This network considers both homogeneous and heterogeneous nodes and nodes are capable to aggregate the data.
- The base station (BS) is situated in the middle of the monitoring area and all the links are symmetric.
- Each sensor node calculates the residual energy, node density, and distance, and each sensor node is capable of self-organizing capabilities.
- All the sensors have similar capabilities, limited computational power, and memory.

In this section, we also discuss a two-tier heterogeneity model where N is the total number of nodes in the network. The energies of level-1 and level-2 nodes are denoted as E_1 and E_2, respectively, that must satisfy the inequality $E_1 < E_2$ and their numbers are denoted as N_1, N_2 respectively, that must satisfy the inequality $N_1 > N_2$. The total energy of the network is determined as follows:

$$E_T = \theta * N * E_1 + (1 - \theta) * N * E_2 \tag{1}$$

where, θ is a model parameter.

Level-1 Heterogeneity For $\theta = 1$, the defined network has only level-1 nodes, that is, all have the same energy and, in this case, the total network energy is given by

$$E_{\text{level}-1} = N * E_1 \tag{2}$$

The level-1 heterogeneity consists of only a single type of nodes which means all the sensor nodes are having the same amount of energy. So, this network is also called *homogeneous networks*. E_1 is the initial energy of the networks.

Level-2 Heterogeneity The level 2 consists of two types of sensor nodes, namely, level-1 and level-2 nodes. The total energy of the network defined in Eq. (1) given by

$$E_{\text{level}-2} = \theta * N * E_1 + (1 - \theta) * N * E_2 \tag{3}$$

The numbers of level-1 and level-2 nodes, denoted by N_1 and N_2, respectively, are given as follows:

$$\left.\begin{array}{l} N_1 = N * \theta \\ N_2 = N * (1 - \theta) \end{array}\right\} \tag{4}$$

Energy of the level-1 and level-2 sensor nodes is given as E_1 and $E_2 = E_1 * (1 + \omega)$, respectively. Let us consider the value of constants θ and ω are 0.4 and 0.25, respectively.

Here, we also discuss a radio dissipation energy model to calculate the transmitting and receiving energy by the sensor nodes during sensing, transmission, and another computational process. The energy depletion for transmitting L-bit message over the short (E_{TXS}) and long (E_{TXL}) distance d is given as follows [3]:

$$E_{\text{TXS}} = L * \epsilon_{\text{elec}} + L * \epsilon_{\text{fs}} * d^2 \qquad \text{if } d \leq d_0 \tag{5}$$

$$E_{\text{TXL}} = L * \epsilon_{\text{elec}} + L * \epsilon_{\text{mp}} * d^4 \qquad \text{if } d > d_0 \tag{6}$$

where ϵ_{elec}, ϵ_{fs}, and ϵ_{mp} are the energy dissipated and d_0 is the distance threshold between a sensor, and the BS as given as follows:

$$d_0 = \sqrt{\frac{\epsilon_{\text{fs}}}{\epsilon_{\text{mp}}}} \tag{7}$$

The total energy consumed by the transmitter in digital coding and modulation is defined in the first term of (5) and (6). The energies spent in receiving (E_{Rx}) and in sensing (E_{Sx}) are given in (8) and (9) as follows:

$$E_{\text{Rx}} = L * \epsilon_{\text{elec}} \tag{8}$$

$$E_{Sx} = L * \in_{elec} \tag{9}$$

In the next section, we will discuss the fuzzy logic system which helps in electing the cluster heads for load balancing.

4 Fuzzy Logic System for the Proposed Protocol

Fuzzy logic system (FIS) is used to select the cluster head and their cluster formation process. This section discusses the fuzzy rule set for the election of cluster heads based on different parameters, namely, residual energy, node density, and distance between a node and base station. The fuzzy logic system consists of four major steps namely, fuzzifier, rule base, fuzzy interface engine, and fuzzified [11]. The fuzzifier accepts crisp input as a crisp number, that is, residual energy, node density, and distance between a node and base station and transforms these values into the fuzzy set by applying functions like trapezoidal, Gaussian, and triangular-shaped. The fuzzy rule base contains various IF-THEN procedures to decide the fuzzy outcome by using a different fuzzy operator like AND/OR. The fuzzy interface engine or aggregation is the collective output of all rules, that is, the maximum values to generate the aggregated fuzzy set. Defuzzification receives inputs from the fuzzy interface and converts them into crisp values as an output value. Defuzzifier procedures used the centroid method for attainment the crisp value also known as the final output probability. In this work, we used Mamdani model for obtaining the output probability in terms of crisp value. The Mamdani model is a simple, easy, and most widely used model in different applications [11]. The centroid defuzzifier function is computed as follows:

$$\text{Centroid function} = \frac{\int \Psi_\eta(x) * x\, dx}{\int \Psi_\eta(x) dx} \tag{10}$$

where, $\Psi_\eta(x)$ signifies the membership function of set η for domain value x. Here, WSNs deliver three input parameters to the FIS, namely, residual energy, node density, and distance, and each input parameter contains three membership functions as indicated in Table 1. The output consists of seven membership functions in terms of probability generated by FIS as shown in Table 2. The input variable membership functions consist of two half trapezoidal and one Gaussian shapes, and output membership functions consist of two half trapezoidal and five Gaussian

Table 1 Output in terms of probability generated by FIS

Output value	Membership functions		
Residual energy	Low (0)	Medium (1)	High (2)
Node density	Sparsely (0)	Medium (1)	Densely (2)
Distance	Near (0)	Medium (1)	Far (2)

shapes as exposed in Fig. 1a–d, respectively. Table 3 indicates the fuzzification relationship between different input and output variables. We have calculated the outcome probability value for deciding the cluster head based on the three input parameters for a particular round. The maximum probability value helps to finalize the cluster head out of deployed sensor nodes in the monitoring area for the current round. The following formula is given for calculating the outcome probability:

Table 2 Output in terms of probability generated by FIS

Output value	Membership functions
Probability	Very weak (0), weak (1), lower medium (2), medium (3), higher medium (4), strong (5), very strong (6)

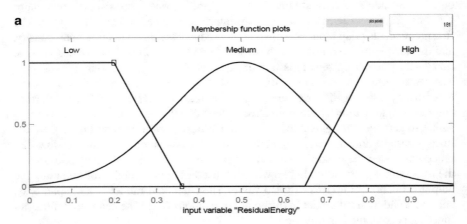

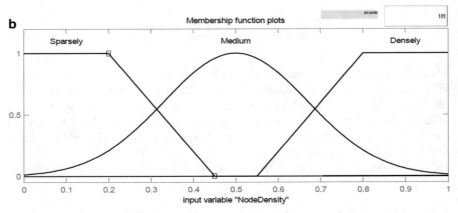

Fig. 1 Membership functions plot corresponding to different inputs and output variables. (**a**) Membership function plot for residual energy input parameter. (**b**) Membership function plot for node density input parameter. (**c**) Membership function plot for distance input parameter. (**d**) Membership function plot for the probability output parameter

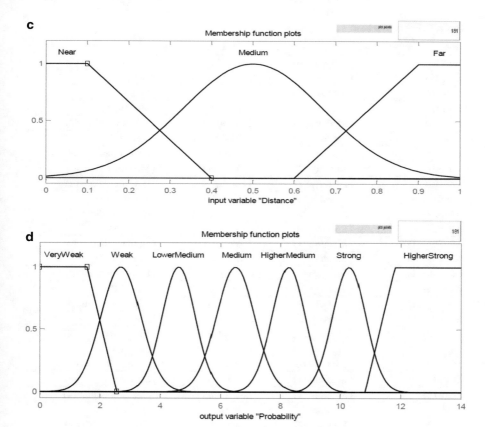

Fig. 1 (continued)

$$\text{Probability} = \frac{w_1 * L_{re} + w_2 * L_{nd} + w_3 * (M_d - L_d)}{w_1 * M_{re} + w_2 * M_{nd} + w_3 * M_d} \qquad (11)$$

w_1, w_2, and w_3 are the weights for the fuzzy input variables; initially, all are considered 1. L_{re}, L_{nd}, and L_d indicate the current level vales of the residual energy, node density, and distance, respectively. M_{re}, M_{nd}, and M_d represent the maximum level value of the residual energy, node density, and distance, respectively.

In the next section, we will discuss the data aggregation process of the data which are collected by the respective cluster head and intermediate cluster heads which are used to forward the data to the base station.

Table 3 Fuzzy rule base

Residual energy	Node density	Distance	Probability
Low (0)	Sparsely (0)	Near(0)	Lower medium (2)
Low (0)	Sparsely (0)	Medium (1)	Weak (1)
Low (0)	Sparsely (0)	Far (2)	Very weak (0)
Low (0)	Medium (1)	Near (0)	Medium (3)
Low (0)	Medium (1)	Medium (1)	Lower medium (2)
Low (0)	Medium (1)	Far (2)	Weak (1)
Low (0)	Densely (2)	Near (0)	Higher medium (4)
Low (0)	Densely (2)	Medium (1)	Medium (3)
Low (0)	Densely (2)	Far (2)	Lower medium (2)
Medium (1)	Sparsely (0)	Near (0)	Medium (3)
Medium (1)	Sparsely (0)	Medium (1)	Lower medium (2)
Medium (1)	Sparsely (0)	Far (2)	Weak (1)
Medium (1)	Medium (1)	Near (0)	Higher medium (4)
Medium (1)	Medium (1)	Medium (1)	Medium (3)
Medium (1)	Medium (1)	Far (2)	Lower medium (2)
Medium (1)	Densely (2)	Near (0)	Strong (5)
Medium (1)	Densely (2)	Medium (1)	Higher medium (4)
Medium (1)	Densely (2)	Far (2)	Medium (3)
High (2)	Sparsely (0)	Near (0)	Higher medium (4)
High (2)	Sparsely (0)	Medium (1)	Medium (3)
High (2)	Sparsely (0)	Far (2)	Lower medium (2)
High (2)	Medium (1)	Near (0)	Strong (5)
High (2)	Medium (1)	Medium (1)	Higher medium (4)
High (2)	Medium (1)	Far (2)	Medium (3)
High (2)	Densely (2)	Near (0)	Very strong (6)
High (2)	Densely (2)	Medium (1)	Strong (5)
High (2)	Densely (2)	Far (2)	Higher medium (4)

5   Proposed Data Collection and Aggregation Process

In this subsection, we will discuss the data collection and data aggregation process
for the stable election protocol.

5.1   Chain-Based Data Gathering and Transmission Process
for Intracluster and Intercluster Communication

The description of the working process of the chain construction approach for
intracluster communication is given as follows.

Step1 Calculate the distance of each sensor from the cluster head and distance among the nodes in the cluster or distance between each cluster head and the BS in the network.

Step2 Select the furthest node from the cluster head/BS which will be the first sensor/cluster head node of the intra/inter communication chaining process in the clusters/networks.

Step3 Furthest/cluster head node starts to choose the next sensor/cluster head node to construct the chain by considering the greedy approach. This approach considers the minimum distant sensor/cluster head node as the next node from the furthest node. Similarly, the next minimum distant sensor/cluster head node is chosen. This process continues until the next minimum distant node/cluster head is the cluster head/networks.

Step4 After forming the first chain in the cluster/networks, if any sensor/cluster head node is not connected with the respective cluster head/base station, then the construction of a new chain will start by considering the same process as defined in Step 3 in the cluster/networks distinctly.

Step5 When the energy of a sensor/cluster head becomes zero, it means the sensor/cluster head dies in the chain and the chain is again reconstructed to avoid the dead node/cluster head.

Step6 Repeat Step 1 to Step 5 for all the possible clusters.

5.2 Data Aggregation Process

The data aggregation algorithm helps in the elimination of duplicate data which is one of the identical activities sensed by the two or more sensors. The removal of duplicate data packets in both intercluster communication and intracluster communication supports in reducing the transmission cost of the networks. Let δ be the number of nodes in a cluster (α) that generates the data packets α_{p1}, α_{p2}, α_{p3}, ..., $\alpha_{p\delta}$,and D_{A1}, D_{A2}, and D_{A3} denote the average, sum, and some of the aggregated data packets, respectively.

The complete process of the data aggregation algorithm is given as follows:

Input: Set of Clusters $C = \{\alpha, \beta, \gamma ....\}$ and base station (BS)
Output: Aggregate data packets at Cluster heads (CH) or BS

Begin
    for every cluster $C = \{\alpha, \beta, \gamma ....\}$
        if $(\alpha_{p1} = \alpha_{p2} = \alpha_{p3} = \cdots = \alpha_{p\delta})$ // all the sensors generated the exact same
    data packets

$$D_{A1} = \left\{ \left(\frac{\alpha_{p1}}{2^{q-1}} \right) + \left(\frac{\alpha_{p2}}{2^{q-1}} \right) + \left(\frac{\alpha_{p3}}{2^{q-2}} \right) + \left(\frac{\alpha_{p4}}{2^{q-3}} \right) + \cdots + \left(\frac{\alpha_{p\delta}}{2} \right) \right\}$$
        // q is number of nodes generated same data packets

else if $(\beta_{p1} \neq \beta_{p2} \neq \beta_{p3} \neq \cdots \neq \beta_{p\delta})$ // *all the sensors generated the different data packets*

$$D_{A2} = \beta_{p1} + \beta_{p2} + \beta_{p3} + \cdots + \beta_{p\delta} \text{ // sum of the data packets}$$

else if $(\gamma_{p1} = \gamma_{p4} = \cdots = \gamma_{p\delta-1}) \neq \gamma_{p2} \neq \gamma_{p3} \neq \gamma_{p\delta}$ // *some sensors generated the exact same data packets and some sensor generated the different data packets*

$$D_{A3} = D_{A1} + D_{A2}$$
$$D_{A3} = \left\{ \left(\frac{\gamma_{p1}}{2^{q-1}} \right) + \left(\frac{\gamma_{p4}}{2^{q-1}} \right) + \left(\frac{\gamma_{p5}}{2^{q-2}} \right) + \left(\frac{\gamma_{p6}}{2^{q-3}} \right) + \cdots + \left(\frac{\gamma_{p\delta-1}}{2} \right) \right\} + \{ (\gamma_{p2} + \gamma_{p3} + \gamma_{p\delta}) \}$$

    end if
  end for
End

6 Simulation Results and Discussions

In this section, the simulation results of the proposed schemes are compared with the existing AZ-SEP [14] and original SEP [15] methods by considering alive and dead nodes per round, stability period, energy consumption per round, and throughput. The stable election protocol (SEP) elects cluster heads using the residual energy of each node [15]. The proposed method is implemented a data aggregation process for removing the redundant data, a chain-based data gathering and transmission process for intracluster communication and intercluster communication. All the proposed implementations are applied to both homogeneous and heterogeneous networks to evaluate the performance. The proposed networks are implemented in MATLAB 2014a environment using 100 numbers of sensor nodes. Initially, all the nodes equipped with 0.5 J initial energy in case of homogeneous networks and BS are fixed at the middle of the monitoring area. The energy consumption to run the transmitter or receiver circuit (ϵ_{elec}), by the amplifier to transmit signal at a shorter distance (ϵ_{fs}), and by the amplifier to transmit at longer distance (ϵ_{mp}) is 50 nJ/bit, 10 pJ/bit/m^2, 0.0013 pJ/bit/m^4, respectively. The message size (L), cluster radius (R), and threshold distance (d_0) are 4000 bits, 25 m, and 75 m, respectively. For each experiment, we have taken 25 simulations using randomly sensor deployment and finally, the average of those 25 simulations is considered as the outcome of the results. Two categories are considered for evaluation of the proposed and existing methods, namely, performance evaluation for homogeneous and heterogeneous networks. In the next subsection, we will discuss the simulation results for homogeneous networks.

6.1 Performance Evaluation for Homogeneous Networks

In this subsection, a comparative result analysis of the original SEP [15], AZ-SEP [14], and the proposed method is discussed for a homogeneous network where all the nodes have the same amount of network energy by considering alive and dead nodes per round, stability period, energy consumption per round, and throughput.

Figure 2 illustrates the number of alive nodes in respect of the number of rounds for original SEP [15], AZ-SEP [14], and the proposed method. It is observed that the proposed method covers 2298 rounds, whereas AZ-SEP [14] and original SEP [15] cover 1628 and 1358 rounds, respectively, before depleting complete energy of all the nodes. Thus, AZ-SEP [14] and the proposed method increase 19.88%, and 69.21% in the network lifetime in respect of original SEP [15], respectively. The proposed method helps in increment in the network lifetime because node density and other parameters reduce the communication cost between the sensor nodes and base station. Moreover, the nodes in the AZ-SEP [14] and proposed method die leisurely than the original SEP [15] because it chooses the head nodes proficiently which helps in extending the lifetime of the network. It is also evident from Fig. 2 the sustainable period of the AZ-SEP [14], and proposed method for first node dead significantly exceeds by 33.02%, and 78.02%, as the comparison with the original SEP [15], respectively. Figure 3 illustrates the number of dead nodes per rounds for original SEP [15], AZ-SEP [14], and the proposed method. The average energy dissipation of the original SEP [15], AZ-SEP [14], and proposed method in respect of the number of rounds for homogeneous networks is shown in Fig. 4. The initial energy of the network is considered as 50 J. The proposed method outperforms as

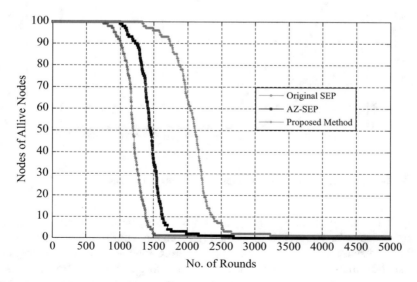

Fig. 2 The number of alive sensor nodes with respect to number of rounds

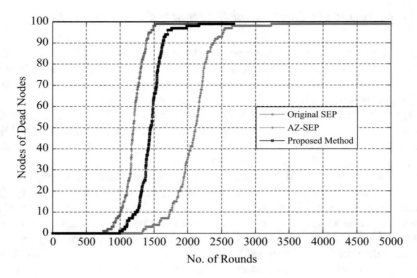

Fig. 3 The number of dead sensor nodes with respect to number of rounds

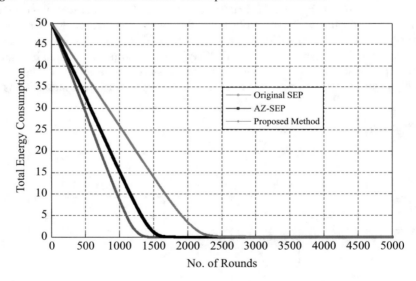

Fig. 4 The total energy consumption with respect to number of rounds

compared to the original SEP [15] and AZ-SEP [14] because it covers a greater number of rounds. The results show the higher energy consumption in the original SEP [15] because it does not consider any chaining approach with data aggregation for data transmission between the nearby sensor nodes and base station. Moreover, the proposed method considers both intracluster communication and intercluster communication by chaining and data aggregation approach which preserves energy

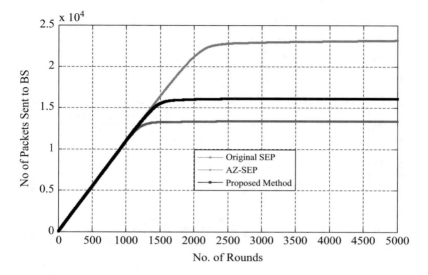

Fig. 5 The number of packets sent to BS with respect to number of rounds

of the sensors in an effective manner. Thus, it reduces the computational energy cost of the head nodes in the networks.

The number of packets received by the base station in respect of the number of rounds till the networks alive using original SEP [15], AZ-SEP [14], and proposed method for the homogeneous networks are shown in Fig. 5. The number of packets sent by the proposed method, AZ-SEP, and original SEP is 2.28×10^{-4}, 1.61×10^{-4}, and 1.32×10^{-4}, respectively. It is observed that the proposed method sent data packets at a higher rate to the base station as associated with the original SEP [15] and AZ-SEP [14]. However, the proposed method produces the highest number of packets because the nodes in the proposed method remain alive more in terms of the number of rounds.

6.2   Performance Evaluation for Heterogeneous Networks

In this subsection, a comparative result analysis of the original SEP [15], AZ-SEP [14], and the proposed method is discussed for the heterogeneous network where nodes have a different amount of network energy by considering alive and dead nodes per round, stability period, energy consumption per round, and throughput.

Figure 6 illustrates the number of alive nodes in respect of the number of rounds for original SEP [15], AZ-SEP [14], and the proposed method. It is observed that the proposed method covers 2945 rounds, whereas AZ-SEP [14] and original SEP [15] cover 2043 and 1698 rounds, respectively, before depleting complete energy of

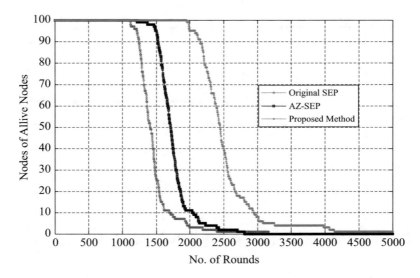

Fig. 6 The number of alive sensor nodes with respect to number of rounds

all the nodes. Thus, AZ-SEP [14] and the proposed method increase 20.31%, and 73.43% in the network lifetime in respect of original SEP [15], respectively. The proposed method helps in increment in the network lifetime because node density and other parameters reduce the communication cost between the sensor nodes and base station. Moreover, the nodes in the AZ-SEP [14] and proposed method die leisurely compared to the original SEP [15] because it chooses the head nodes proficiently which helps in extending the lifetime of the network. It is also evident from Fig. 6 the sustainable period of the AZ-SEP [14], and proposed method for first node dead significantly exceeds by 9.10%, and 76.31%, as a comparison with the original SEP [15]. Figure 7 illustrates the number of dead nodes per rounds for original SEP [15], AZ-SEP [14], and the proposed method.

The average energy dissipation of the original SEP [15], AZ-SEP [14], and proposed method in respect of the number of rounds for heterogeneous networks is shown in Fig. 8. The initial energy of the network is considered as 60 J. The proposed method outperforms as compared to the original SEP [15] and AZ-SEP [14] because it covers a greater number of rounds. The results show higher energy consumption in the original SEP [15] because it does not consider any chaining approach with data aggregation for data transmission between the nearby sensor nodes and base station. Moreover, the proposed method considers both intracluster communication and intercluster communication by chaining and data aggregation approach which preserves the energy of the sensors in an effective manner. Thus, it reduces the computational energy cost of the head nodes in the networks.

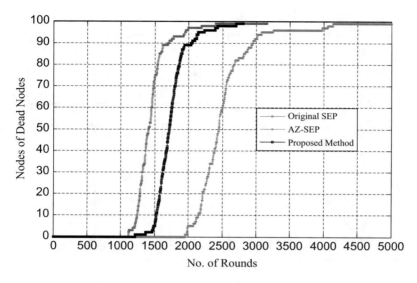

Fig. 7 The number of dead sensor nodes with respect to number of rounds

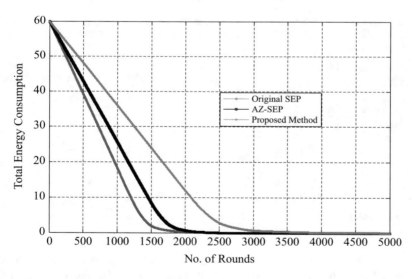

Fig. 8 The total energy consumption with respect to number of rounds

The number of packets received by the base station in respect of the number of rounds till the networks alive using original SEP [15], AZ-SEP [14], and proposed method for the heterogeneous networks is shown Fig. 9.

The number of packets sent by the proposed method, AZ-SEP, and original SEP is 2.89×10^{-4}, 1.98×10^{-4}, 1.67×10^{-4}, respectively. It is observed

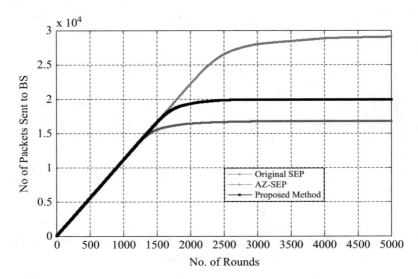

Fig. 9 The number of packets sent to BS with respect to number of rounds

Table 4 Comparative analysis in terms of network lifetime, energy consumption, and throughput for original SEP [15], AZ-SEP [14], and proposed method in case of both homogeneous and heterogeneous networks

Protocols	Network lifetime			Energy consumption (J)	Throughput	% increment in network lifetime
	FND	HND	LND			
Homogeneous networks						
Original SEP	745	1195	1358	50	1.32×10^{-4}	–
AZ-SEP	991	1455	1628	50	1.61×10^{-4}	19.88
Proposed method	1327	2109	2298	50	2.28×10^{-4}	69.21
2-level heterogeneous networks						
Original SEP	1110	1415	1698	60	1.67×10^{-4}	–
AZ-SEP	1211	1725	2043	60	1.98×10^{-4}	20.31
Proposed method	1957	2444	2945	60	2.89×10^{-4}	73.43

that the proposed method sent data packets at a higher rate to the base station as associated with the original SEP [15] and AZ-SEP [14]. However, the proposed method produces the highest number of packets because the nodes in the proposed method remain alive more in terms of a number of rounds.

Table 4 shows the comparative analysis in terms of network lifetime, energy consumption, throughput, and percentage increment in network lifetime for original SEP [15], AZ-SEP [14], and proposed method in case of both homogeneous and heterogeneous networks. In the case of homogeneous networks, throughput is 2.28×10^{-4}, 1.61×10^{-4}, and 1.32×10^{-4} for the proposed method, AZ-SEP [14] and original SEP [15], respectively, but for heterogeneous networks, throughput is

2.89×10^{-4}, 1.98×10^{-4}, 1.67×10^{-4} for the proposed method, AZ-SEP [14], and original SEP [15], respectively. The proposed methods transmitted a greater number of data packets because the lifetime of the proposed networks is better than the existing ones. The increment in the network lifespan of the AZ-SEP [14] and proposed method is 69.21% and 19.88% as compared with the original SEP [15] without any increment in the network lifespan, respectively, for homogeneous networks; similarly, increment in the network lifespan of the AZ-SEP [14] and proposed method is 73.43% and 20.31% as compared with the original SEP [15] without any increment in the network lifespan, respectively, for heterogeneous networks.

7   Conclusion

In this chapter, a clustering-based optimized stable election protocol for prolonging the network lifetime in WSNs is proposed. The proposed technique considered both homogeneous and heterogeneous networks for original SEP [15], AZ-SEP [14], and the proposed method. The proposed method used a fuzzy-based clustering technique for cluster head election procedure and chaining approach with data aggregation for efficient data collection. It provides a sustainable region in the network execution for both networks because the selection of higher residual energy nodes for cluster head selection is efficient. The simulation results demonstrate that lifetime for homogeneous and heterogeneous networks is increased by 19.88%, 69.21%, and 20.31%, 73.43% for 50 and 60 J network energy in case of AZ-DEEC and proposed a method with respect of the original SEP [15]. The proposed method performs best among the original SEP [15] and AZ-SEP [14].

References

1. Singh, S., Chand, S., & Kumar, B. (2013). Performance investigation of heterogeneous algorithms in WSNs. In *3rd IEEE International Advance Computing Conference (IACC)*, pp. 1051–1054.
2. Singh, Y., Singh, S., & Kumar, R. (2012). A distributed energy-efficient target tracking protocol for three level heterogeneous sensor networks. *International Journal of Computer Applications, 51*, 31–36.
3. Heinzelman, W. R., Chandrakasan, A. P., & Balakrishnan, H. (2002). An application-specific protocol architecture for wireless microsensor networks. *IEEE Transactions on Wireless Communications, 1*, 660–670.
4. Lindsey, S., Raghavendra, C. S., & Sivalingam, K. M. (2002). Data gathering algorithms in sensor networks using energy metrics. *IEEE Transactions on Parallel and Distributed Systems, 13*, 924–935.
5. Singh, S., & Malik, A. (2017). hetSEP: Heterogeneous SEP protocol for increasing lifetime in WSNs. *Journal of Information and Optimization Sciences, 38*, 721–743.

6. Qing, L., Zhu, Q., & Wang, M. (2016). Design of a distributed energy-efficient clustering algorithm for heterogeneous wireless sensor networks. *Computer Communications, 29*, 2230–2237.
7. Singh, S., Malik, A., & Kumar, R. (2017). Energy efficient heterogeneous DEEC protocol for enhancing lifetime in WSNs. *Engineering Science and Technology, an International Journal, 20*, 345–353.
8. Maheswari, D. U., & Sudha, S. (2018). Node degree based energy efficient two-level clustering for wireless sensor networks. *Wireless Personal Communications, 104*, 1209–1225.
9. Chand, S., Singh, S., & Kumar, B. (2014). Heterogeneous HEED protocol for wireless sensor networks. *Wireless Personal Communications, 77*, 2117–2139.
10. Singh, S., Chand, S., & Kumar, B. (2016). Energy efficient clustering protocol using fuzzy logic for heterogeneous WSNs. *Wireless Personal Communications, 86*, 451–475.
11. Singh, S., Chand, S., & Kumar, B. (2017). Multilevel heterogeneous network model for wireless sensor networks. *Telecommunication Systems, 64*, 259–277.
12. Singh, S., Chand, S., & Kumar, B. (2014). An energy efficient clustering protocol with fuzzy logic for WSNs. In *5th International Conference-Confluence the Next Generation Information Technology Summit*, pp. 427–431.
13. Faisal, S., Javaid, N., Javaid, A., Khan, M. A., Bouk, S. H., & Khan, Z. A. (2013). Z-SEP: Zonal-stable election protocol for wireless sensor networks. *Journal of Basic and Applied Scientific Research, 3*(5), 132–139.
14. Khan, F. A., Khan, M., Asif, M., Khalid, A., & Haq, I. U. (2019). Hybrid and multi-hop advanced zonal-stable election protocol for wireless sensor networks. *IEEE Access, 7*, 25334–25346.
15. Smaragdakis, G., Matta, I., & Bestavros, A. (2004). *SEP: A stable election protocol for clustered heterogeneous wireless sensor networks* (Technical Report BUCS-TR-2004-022). Boston University Computer Science Department, pp. 1–11.
16. Singh, S., Chand, S., & Kumar, B. (2013). 3-Tier heterogeneous network model for increasing lifetime in three dimensional WSNs. *QSHINE, 2013*, 238–247.
17. Singh, S., & Malik, A. (2017). hetDEEC: Heterogeneous DEEC protocol for prolonging lifetime in wireless sensor networks. *Journal of Information and Optimization Sciences, 38*, 699–720.
18. Chand, S., Singh, S., & Kumar, B. (2013). hetADEEPS: ADEEPS for heterogeneous wireless sensor networks. *International Journal of Future Generation Communication and Networking, 6*, 21–32.
19. Chand, S., Singh, S., & Kumar, B. (2013). 3-Level heterogeneity model for wireless sensor networks. *International Journal of Computer Network and Information Security, 5*, 40–47.
20. Manju, Chand, S., & Kumar, B. (2018). Target coverage heuristic based on learning automata in wireless sensor networks. *IET Wireless Sensor Systems, 8*, 109–115.
21. Manju, Chand, S., & Kumar, B. (2016). Maximising network lifetime for target coverage problem in wireless sensor networks. *IET Wireless Sensor Systems, 6*, 192–197.
22. Mehra, P. S., Doja, M. N., & Bashir, A. (2019). Enhanced clustering algorithm based on fuzzy logic (E-CAFL) for WSN E-CAFL for homogeneous WSN. *Scalable Computing: Practice and Experience, 20*, 41–54.

Feature Selection Is Important: State-of-the-Art Methods and Application Domains of Feature Selection on High-Dimensional Data

G. Manikandan and S. Abirami

1 Introduction

Since high-dimensional data could provide ample information, it is hard to establish a precise prediction model along with the growing dimensionality and scale of dataset. High-dimensional data are the data that have more number of variables or attributes. With recent advances in data acquisition and storage technology, high-dimensional data extensively occur in nature, finance, industry, biomedicine, and several other fields, which contain complex nonlinear relationship among multiple features [1, 2]. There are three categories of attributes in prediction model: relevant, redundant, and irrelevant. The relevant feature is highly correlated with the target, whereas redundant features are correlated with each other; in case of irrelevant feature, they do not have any significant information on target.

In order to select the relevant features by reducing the redundant and irrelevant feature selection provides the way with increased accuracy. Feature selection varies with supervised learning, unsupervised learning, and semi-supervised learning. In supervised learning feature selection, it uses the label/class data to select the features, that is, it selects the feature based on class information. In unsupervised learning feature selection, it selects the feature without class information, based on the feature–feature relation that eliminates irrelevant features. The main demerits of this method are that it ignores the correlation between the features and class information and also sometimes, it ignores the correlation between features [3].

G. Manikandan (✉)
Department of Computer Science and Engineering, College of Engineering Guindy,
Anna University, Chennai, Tamil Nadu, India

S. Abirami
Department of Information Science and Technology, College of Engineering Guindy,
Anna University, Chennai, Tamil Nadu, India

© Springer Nature Switzerland AG 2021
R. Kumar, S. Paiva (eds.), *Applications in Ubiquitous Computing*, EAI/Springer
Innovations in Communication and Computing,
https://doi.org/10.1007/978-3-030-35280-6_9

Semi-supervised feature selection is the combination of supervised and unsupervised feature selections; it uses and selects the feature with and without class information.

The feature selection process consists of three main stages: in the first stage, it identifies the search direction by taking any one of the randomly selected features or it may be an empty set. After that, it selects a new feature and adds into the feature set with "n" number of iterations till the final optimal set [2]. This type-selecting feature from the empty set to the optimal set is called forward search. In addition to this, the process of selecting feature from full set and by eliminating the irrelevant feature in each iteration in order to get the final subset is called backward elimination. Sometimes, the search direction performs both forward selection and backward elimination methods to select the predominant features in the each of the iteration, which is called bidirectional search method. Based on the properties of the dataset, the search direction has to be selected by the researchers to perform better feature selection. The second stage of the feature selection process is to identify the search strategies. There are three main strategies: exponential, random search, and sequential search. Determining the evaluation criteria is the last stage of feature selection. Filter, wrapper, hybrid, and embedded are the best evaluation criteria of the feature selection process [4].

In the medical domain, gene expression may have more than thousands of genes among the vast amounts of genes and many of the genes will be completely irrelevant for classifying the purpose and also it leads to overfitting. Due to the presence of overfitting, it needs more computational time and processing power to classify the data with high accuracy. Most of gene expression data contain irrelevant, redundant information; here, the feature selection plays an important role to select the highly informative genes from a vast number of genes in order to reduce the computational cost with a high classification accuracy. These types of microarray data analysis help to find out deadly diseases like cancer early. Mostly humans may suffer 200 types of cancer; for this, microarray analysis [3, 4] was used to detect the cancers early. Breast cancer, colon cancer, leukemia, prostate cancer, and lung cancer are some types of cancer which the human beings encounter.

Dimensionality reduction can be done by two ways, namely feature selection and feature extraction, where feature selection selects the predominant feature from the vast data without modification of the originality of the data, that is, it is a subset of the whole data, whereas the feature extraction converts the original data into some form based on the application of linear or nonlinear transformation to reduce the data [3]. The static feature selection (FS) and streaming feature selection are the two categories of FS; the original data do not change over time, whereas the new feature may be added to original data in streaming feature selection. In this chapter, we have focused on only feature selection methods such as filter, wrapper, embedded, ensemble, and hybrid methods which are shown in Fig. 1.

In this chapter, Sect. 1 describes the introduction to feature selection, process of feature selection, and objectives of feature selection, whereas Sect. 2 summarizes the various related works with respect to filter, wrapper, embedded, and hybrid methods. Section 3 provides various application fields such as microarray analysis,

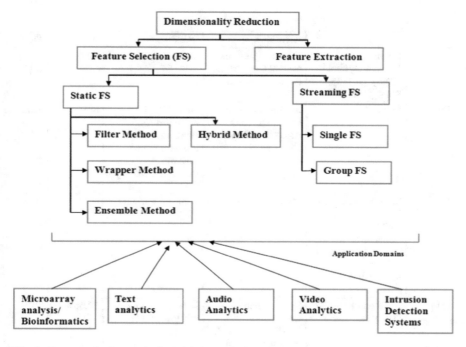

Fig. 1 Feature selection methods and their applications

text analysis, video analytics, audio analytics, and intrusion detection systems; discussion among various feature selection methods is given in Sect. 4. Section 5 concludes the chapter with the issues relating to selecting the optimal feature subset and future directions in dimensionality reduction.

2 Related Works

2.1 Filter Methods

In filter methods, there are various ranking approaches used as the principle criteria for feature selection. Ranking methods assign ranks to each feature based on some intrinsic and statistical properties, and an appropriate ranking method is selected and used to rank the features; after this, based on the threshold value, the feature can be selected. Here, the threshold value is selected by applying their own proposed algorithms or by using trial-and-error method. Filter methods select features based on the internal characteristics of the features. They do not involve any classifiers to select the features. Table 1 presents the recent important techniques based on the filter method.

Table 1 Recent techniques based on filter method

Ref. no.	Dataset	Algorithm	Classifier	Accuracy	Year
[5]	Colon, Central Nervous System (CNS), GLI_85, SMK_CAN_187	Whale Algorithm, Mutual Congestion, Forward Feature Selection Method	SVM, NB, DT	80%	2019
[6]	Emails	Information gain, Latent Dirichlet Allocation, Topic Guessing	NB, SVM, C4.5, Adaboost C4.5, Bagging C4.5, Random Forest, Logistic Regression, Rough Sets	88%	2019
[7]	Ionosphere, Breast Cancer Wisconsin (BCW), Connectionist Bench, Iris, Statlog (Vehicle Silhouettes), Parkinson	Modified-BPSO	K-means Clusters	–	2019
[8]	NIST and RIMES Databases	Symmetrical Uncertainty, Chi-Square, Relief, Information gain, Gain Ratio, Correlation-based Feature Selection, Consistency Criterion	K-NN, Bagging and Random Forest	72.66%	2019
[9]	Simulated and Industrial datasets	Wide Spectrum feature selection for regression (WiSe)	Forward Stepwise regression (FSR), Selector operator (LASSO), Partial Least Squares (PLS) and Least Absolute Shrinkage	–	2019

2.2 Wrapper Method

The wrapper method selects the features with the help of machine learning algorithms by knowing the classification accuracy and error rate. This method aims to minimize the classification error and to improve the classification performance. In general, it gives better accuracy than the filter method because it tunes the model and selects the feature based on the learning algorithm. The main disadvantages of this method are that it consumes more computational time, and also the classifier performance may vary across the different classifiers. Mostly, SVM, Naïve Bayes,

Table 2 Recent techniques based on wrapper method

Ref. no.	Dataset	Algorithm	Classifier	Accuracy	Year
[10]	Colon, Central Nervous System (CNS), GLI_85, SMK_CAN_187	Binary Gross hopper Optimization algorithm	k-Nearest Neighbor	77.04%	2019
[6]	Emails	Multi objective Evolutionary Algorithm	Linear Discriminant Analysis, k-Nearest Neighbor	89.9%	2019
[11]	Breast cancer, BreastEW, Exactly, Exactly2, HeartEW, Lymphography, M-of-n, penglungEW, SonarEW, SpectEW, CongressEW and IonosphereEW datasets	Las Vegas Wrapper	Ensemble learner	97.04%	2019
[12]	Real Motory image dataset	Wrapper-based selection	J48, PART, Adaboost, Random Forest, Naïve Bayes	76.33%	2019
[13]	Kidney disease dataset	Chaotic crow search algorithm (CCSA)	k-Nearest Neighbor	83.8%	2019

and Random Forest algorithms are used as classifiers in the wrapper method. Some of the recent wrapper-based methods and applied datasets with accuracy are presented in Table 2.

2.3 Hybrid Method

The hybrid feature selection method performs the feature selection process by merging or joining two different feature selection methods. For example, we can merge the filter method and wrapper method into one and perform the selection process, but the evaluation criterion has to be the same for the two methods. In the hybrid method, it uses the merits of the two methods for selecting the features and combines the selected features into a single subset. Sometimes, this method uses different evaluation criteria and search strategies for gaining accuracy with less computational time. For example, the filter method is used for selecting the initial

Table 3 Recent techniques based on hybrid method

S. no.	Dataset	Algorithm	Classifier	Accuracy (approx)	Year
[14]	Medical datasets	Roughset, BGWO, BCGWO	K-nearest neighbor	72.6%	2019
[15]	Phishing dataset	Hybrid Ensemble Feature Selection, Cumulative Distribution Function Gradient	Naïve Bayes Random Forest, SVM, C4.5, JRip	94.6%	2019
[16]	Breast Cancer, Zoo, Tic-tac-toe, Vote, Waveform, Wine, CongressEW, Lymphography	BGWOPSO	K-nearest neighbor	93%	2019
[17]	Ionosphere, Automobile, BreastDiagnostic, BreastPrognostic, German, SPECTF Heart HillValley, Ozone Level, Parkinsons and Sonar	MPMDIWOA-Maximum Pearson MaximumDistance Improved Whale Optimization Algorithm	SVM	90.4%	2019
[18]	Irvine repository datasets	Grey Wolf Optimization and Crow Search Algorithm	K-nearest neighbor	90.61%	2019

features by assigning weights; later, the same features will be given to the wrapper method in order to get optimized results. Some of the recent hybrid-based methods and applied datasets with accuracy are presented in Table 3.

2.4 Embedded Method

Embedded method uses the machine algorithms to select the features based on internal optimization of the features. Here, the features are selected based on the predefined function to evaluate the features. Many research studies reveal that the embedded methods are more convenient to select the features with less computational cost. This is because this method does not need continuous evaluation of the feature repetitively and is less prone to overfitting. Ensemble feature selection follows the idea of ensemble learning which uses the aggregated results of various learners. Ensemble feature selection aims to find stable and robust feature subset. In order to find a stable subset, this method employs different types of feature selectors.

Table 4 Recent techniques based on embedded method

S. no.	Dataset	Algorithm	Classifier	Accuracy (approx)	Year
[19]	Mushroom, Thyroid, Diabetes, Liver, Breast Cancer, Heart, CKE, Dermatology, Ionosphere, Tumour data, Audiology, Lymphography, Zoo	Wrapper Method, Bagging	SVM, RF, NB	92%	2019
[20]	Diabetic Electronic Medical Records	GBM – Gradient Boosting Machine with mean rank	–	82%	2019

Finally, results from various feature selectors are aggregated to produce the desired feature subset. Some of the recently proposed embedded-based methods and applied datasets with accuracy are presented in Table 4.

2.5 Feature Selection Based on Fuzzy Logic

Currently, researchers concentrate on computational intelligence-based techniques such as fuzzy rough set-based attribute selection, fuzzy support vector machines (FSVM), high-performance feature selection, fuzzy feature selection, supervised neural networks, unsupervised neural networks, etc., for selecting the best features. Despite feature selection, feature relevance, feature redundancy, and anomaly detection are other issues which are present in the classification of high-dimensional dataset. Apart from this, feature grouping or feature clustering is one of the approaches which are being currently used in most of the application domains to reduce the dimensionality of feature before classification. In order to achieve the abovementioned, we have presented some of the recent fuzzy logic feature selection techniques in Table 5.

2.6 Comparison of Feature Selection Methods

Filter methods select the features without the help of classifiers; hence, it is faster and gives considerable accuracy. Compared to the filter method, it selects the feature with the interaction of the classifier and also provides higher accuracy than filter because of tuning the parameters with the classifiers, but it takes more computation

Table 5 Recent techniques based on fuzzy logic

S. no.	Dataset	Algorithm	Classifier	Accuracy (approx)	Year
[21]	Flash Flood Dataset	FURIA-GA	C4.5 Decision Trees, JRip	89.03%	2019
[22]	Cancer Dataset	TCGA	Fuzzy Rule-based classification	90%	2019
[23]	ALL AGENTS and INBOUND AGENTS	MultiObjective-EvolutionarySearch method with the multi-objective evolutionary algorithm ENORA	Fuzzy Rule-based classification	73.25%	2019
[24]	Diabetes dataset	Fuzzy principal component analysis (FPCA)	FPCA-SVM	71%	2019
[25]	39 Bus drivers data	Adaptive Neuro-Fuzzy Inference System (ANFIS), Particle Swarm Optimization (PSO)	SVM	98.12%	2019

time. Embedded methods select the feature with classifier and give better accuracy compared to the wrapper method, but it does not consider the overfitting problem. The hybrid method is the combination of feature selection methods and it gives better accuracy than the filter method and also gives better computation cost when compared to the wrapper method.

3 Importance of Feature Selection and Application Domains

This section provides the importance of feature selection in various application domains such as text categorization, video analytics, audio analytics, microarray data analysis, bioinformatics, instruction detection techniques, and streaming data analysis. Also, this section discusses the various recent feature selection techniques with respective application domains.

3.1 Importance of Feature Selection in Text Categorization

Due to the fast growth of the text-based content on WWW, it has led to the problem of categorizing the content based on a particular context. In text mining, identifying the common pattern among the text documents is a difficult task because most of the text data contain irrelevant and redundant data. The application domain includes sentiment analysis, bioinformatics, movie/product recommendation, web spam detection, clinical data analysis, etc. In particular, dimensionality reduction by taking into consideration both feature extraction and feature selection plays an important role in text categorization. Feature selection selects the feature subset from the vast amount of dataset without modifying the original data, whereas feature extraction transforms or combines the features based on certain models for reducing and selecting the predominant features. Alternatively, the feature extraction technique is used to extract the core feature of the text data, whereas feature selection is used to select the predominant features. Many statistical and machine learning-based text categorization techniques have been proposed by researchers in order to extract the core features from the vast amount of text data. However, there are problems and challenges that exist still: to select the feature and in classifying the text document automatically.

Authors from the study cited herein [26] mentioned two major problems which are currently facing the researchers in the field of text categorization: the first one is huge numbers of dimensionality in the data and the second one is the presence of noisy features in the text document while creating the vector space model through the Bag of Words (BoW) approach. Due to the existence of these problems, the computational complexity becomes high and also leads to problems in the classification accuracy. As mentioned in the previous section, the feature selection method can be classified into filter, wrapper, embedded, hybrid, and ensemble approaches.

Multivariate relative discriminative criterion (MRDC) is one of the effective multivariate filter-based feature selection techniques which has been proposed to select the predominant features from the text data [27]. MRDC is mainly designed for effective feature selection and text classification. This technique consists of three main steps such as preprocessing, feature selection, and evaluation. Preprocessing is common in data mining; the main objective is to perform stop-word removal, stemming, pruning, and term-weighting. As a second step, the authors addressed and proposed the feature selection technique to select the informative feature from the text data with consideration of the two principles' relevancy by employing relative discriminative criterion (RDC) and redundancy by using Pearson correlation. Finally, the subset of the features is selected from the whole feature set. WebKB, Reuters-21578, and 20-Newsgroup datasets are used for evaluating their proposed algorithm, and results are evaluated in terms of precision, recall, and F-Measure. Decision tree and multinomial naïve Bayes and multilayer perceptron were used as the classifier algorithms. Friedman test has been used to prove that their obtained results are statistically significant.

Feature selection in text is classified into four categories: syntactic, semantic, stylistic, and information gain-based methods. The syntactic model focuses on selecting the model automatically by learning the features in the document by separating the subjective expressions from the polarities. This model focuses only on the subjective-based expressions by ignoring the irrelevant features. The semantic-based approach represents the document as the collection of words, where the sentiment of the each word can be predicted based on the linguistic features such as noun, adjectives, and verbs. Stylistic-based text analysis approach focuses on the various semantic functions of the words and phrases based on the usefulness of the features. Feature selection based on information gain selects the features based on the entropy values between the features/variables and class.

3.2 Importance of Feature Selection in Speech Recognition

Speech recognition plays an important role in many application areas such as the medical domain for detecting stress and pain, robot interactions, computer games, cyber forensics, and call centers for predicting the speaker emotions. These emotions can be predicted by applying various pattern recognitions, machine learning, and artificial intelligence algorithms by finding the patterns and classification. In general, the emotion in the speeches varies among different speakers, but it is important to analyze the emotions for classification. Speech emotion recognition is one of the techniques which extract the emotions in the speech signal. Feature extraction, feature selection, and emotion recognitions are the basic steps involved in the speech recognition. Linear Predictor Coefficients (LPC), Mel-Frequency Cepstral Coefficients (MFCC), Linear Predictor Cepstral Coefficients (LPCC) significantly contribute to emotion recognition and are some of the important feature selection methods applied in speech recognition which are mentioned in Kasiprasad Mannepalli et al. [28].

Currently, researchers focus on finding the number of speech features that are used to categorize the emotional content of the speech. For selecting the features, many feature selection algorithms have been developed and used to select the features, as well as to improve the classification performance. Some of the feature selection methods which are used to select the important features are as follows:

(a) Fast correlation-based filter
(b) Forward feature selection (FFS) and backward feature selection (BFS)
(c) Wrapper-based feature selection
(d) Fuzzy-based feature selection
(e) Sequential floating forward selection (SFFS)
(f) Principal component analysis
(g) Least squares (LS) bound
(h) Mutual information (MUTINF)
(i) Minimum redundancy maximum relevance (mRMR)

ReliefF, Symmetrical uncertainty, Fisher score, spectral feature selection (SPEC), Laplacian score, sparse, local feature selection based on scatter separability (LFSBSS), multi-cluster-based feature selection (MCFS), relief, inconsistency criterion, clustering-based feature selection, and ReliefC.

Authors of the study cited herein [29] proposed a novel feature selection technique for speech emotion recognition based on observing the changes in feature subset according to particular emotions. In their experiment analysis, four

Table 6 Feature selection algorithms on speech recognition

Ref. no.	Dataset	Feature selection algorithm	Classifiers used	Accuracy	Year
[29]	EMO-DB, eNTERFACE05, EMOVO SAVEE	A novel feature selection method for speech emotion recognition	SVM, k-NN, MLP	42.60–84.07%	2019
[30]	Tollywood and Bollywood Popular Songs (TBPS).	GAFS – Genetic algorithm-based feature selection	SVM – support vector machines, ANN – Artificial neural networks, RF – random forest	74.69–91.58%	2018
[31]	EMO-DB, RML, eNTERFACE05, and BAUM-1s	DTPM – Discriminant Temporal Pyramid Matching	Deep convolutional neural networks (DCNN) with five layers	EMO-DB (87.31–86.30%) RML (75.34–75.20%) eNTERFACE05 (79.25–79.40%) BAUM-1s (44.61–44.03%)	2019
[32]	eNTERFACE	Emotion recognition using deep learning approach	RBF kernel in the SVM Polynomial kernel in the SVM	99.9%	2019
[33]	SAVEE, Emo-DB, MES and DES	Salient discriminative feature analysis (SDFA)	CNN	71.8% (SAVEE) 57.2% (Emo-DB) 60.4% (DES) 57.8% (MES)	2014

datasets, namely EMO-DB, eNTERFACE05, EMOVO, and SAVEE, were used for the analysis features and for classification, support vector machine, multilayer perceptron, and k-NN classifiers were used. And also, the proposed feature selection methods are compared with the standard methods such as principal component analysis, fast correlation-based feature selection, and sequential forward selection.

Some of the recent speech emotion recognition methods and applied datasets with accuracy are presented in Table 6.

3.3   Importance of Feature Selection in Video Processing

In video analytics area, the feature selection is used in selecting the important features. The application of the video analytics includes surveillance of public data to predict the crime, face recognition, gait recognition, human action recognition, etc. Object tracking and recognition of the robots is one of the trending areas and also vehicle detection from the traffic video avoids the crimes and helps safe driving. Feature selection with Joint $l_{2,1}$-*norm* minimization (FSNM), Minimum-Redundancy–Maximum-Relevance (MRMR), Fisher Score, Fast Correlation-Based Filter (FCBF), $l_{2,1}$-*norm* Manifold (L21), SParse Multinomial Logistic Regression (SBMLR) via Bayesian L1 Regularization [34] are some the standard existing techniques which are used to compare the proposed feature selection technique which is currently being developed.

Mobile video streaming QoE prediction is one the research areas in video analysis, particularly in the video providing services such as YouTube and Netflix. The main aim of the providers is to provide high quality and at considerable operational cost, but there is a trade-off between these two. In general, the prediction of the QoE has been divided into two types, namely continuous time QoE and retrospective QoE. In retrospective QoE, a single score provided by the subjects will describe the overall QoE for the entire video, whereas continuous-time QoE provides the real-time measurement of each subject with the current QoE, which may lead to trigger the current quality of the video. HAS, which is called the HTTP-based adaptive video streaming, has been proposed [35] which selects the features based on the video quality features, number of staling event, etc., and the experimental analysis is conducted on LIVE-Netflix DB and Waterloo DB.

Some of the recent video processing methods and applied datasets with accuracy are presented in Table 7.

3.4   Importance of Feature Selection in Intrusion Detection Systems

Due to recent advancement in network-based technology, the threat of spammers, criminals, and attackers has also been increasing. The total annual financial loss which is caused due to network intrusion was about US$130 million in 2005; now, it will be more than thrice. Intrusion detection is the technique for detecting the intruder's attacks on the networks; these attacks can be detected by signature-based misuse's detection or anomaly-based detection. In the misuse-based detection system, the patterns of attacks are already stored in the network data and database, and if the data are matched with the database, then it is declared as attack. Anomaly detection creates the profile afterward; it analyzes and observes the behavior of the network.

One of the issues in intrusion detection system is accuracy of the classification, since most of the datasets are imbalanced. In this case, feature selection algorithms

Table 7 Feature selection algorithms on video processing

Ref. no.	Dataset	Feature selection algorithm	Classifiers used	Accuracy	Year
[36]	MSR Action3D, MSRDailyActivity3D, Online RGBD Action	Novel method for feature selection based on a Markov blanket combined with the wrapper method	HMM (Hidden Markov Model), DBN (Dynamic Bayesian Network)	91.80% MSRDaily-Activity3D 94.17% Online RGBD Action 97.95% Chalearn LAP 2014 95.23%	2017
[34]	Youtube Kodak HumanEva MIR FLICKR COIL-20 COREL-50	GLocal Structural feature selection with Sparsity (GLSS)	SVM Adaboost KNN	Accuracy for SVM(*NoFS/ Proposed*) Youtube 38.2%/34.3% Kodak 49.3%/45.9% HumanEva 95.7%/83.4% MIR FLICKR 52.7%/46.1% COIL-20 82.2%/68.7% COREL-44.3%/38.2%	2014
[37]	TRECVID 2012 Open videos (OV) YouTube videos	Video Semantic Analysis-based Kernel Locality-Sensitive Discriminative Sparse Representation (KLSDSR)	–	Recognition Rate TRECVID 2012 – 91.20% Open videos (OV) – 89.20% YouTube videos – 90.17%	2019

are used to select the predominant features in order to classify and improve the efficiency of the algorithm. Filter, wrapper, and embedded-based feature selection techniques were used to reduce and eliminate the number of features which were explained in the previous section. The main types of attacks are DoS attack, replay attacks, selective forwarding attack, Sybil attack, Sinkhole attack, Wormhole attack, black hole attack, Jamming attack, false data attack, etc.

KDD Cup 1991 dataset is one the well-known datasets for the intrusion detection system which consists of five million records, and each record consists of 41 nominal and continuous features with the class label (Normal, DOS, Probe, U2R, R2L) which are available in http://kdd.ics.uci.edu/.

The authors from the study cited herein [38] proposed intelligent fuzzy rough set-based feature selection algorithm and temporal classification for intrusion detection system in WSNs for selecting the important attributes in order to predict the attacks and also a Fuzzy Rough set-based Nearest Neighborhood technique (FRNN) is developed for the effective classification of the multiclass data. Their proposed FRNN approach gives better detection accuracy of about 99.87%.

Correlation and interact feature selection (based on symmetrical uncertainty), Random Forest-Backward Elimination Ranking (RF-BER) FS and Random Forest-Forward Selection Ranking RFFSR), Markova blanket model and decision tree analysis in feature selection, NPGA algorithm, NSGA, NGSA-II, GHSOM-pr, Fuzzy Enhanced Support Vector Intrusion Detection model (Fuzzy ESVDF) using Fusion of chi-square feature selection, Latent Dirichlet Allocation (LDA) and genetic algorithm are some of feature selection algorithms which are used in intrusion detection systems.

3.5   Importance of Feature Selection in Microarray Data Analysis

During the last few decades, the advancement in DNA microarray data analysis has created a direction of the research in machine learning, statistical analysis, and bioinformatics. Generally, these types of DNA microarray data are collected from the tissue samples of the persons and based on the differences in the gene expressions, the person will be distinguished by specific tumors. Particularly, these microarray medical datasets consist of tens and thousands of features with less number of instances since with small sample size with large number of features it suffers with classification accuracy and computational time.

Prediction and classification of the cancer-infected genes and normal healthy genes from the microarray data are always a problem in modern society. Most important thing in the prediction is how well the generated or proposed algorithm differentiates those genes because gene dataset usually consists of a lot of noise values and more number of features. From those features, not all the features are useful in classifying the gene as cancer or regular genes. So, an efficient feature selection algorithm is needed to select and extract only informative genes from the vast number of other genes. For solving this issue, many feature selection techniques such as filter, embedded, wrapper, and hybrid methods have been proposed to select the subset of informative features from the high-dimensional dataset.

Feature selection is the preprocessing step to overcome the issues of selecting features from the microarray data. One of the applications of the microarray data analysis is to predict the cancer early since many types of cancer are caused due to the epidemic or genetic changes. Microarray data analysis is one of the established standard tools which are used to identify and analyze the gene data. One of the main functionalities of the microarray data analysis is to monitor the gene expression level from the genome scale, and after the process of genome scale, the experiment results form a matrix called gene expression matrix. The gene expression matrix consists of genes along with persons, where each row represents the persons' sample instances and each column represents the gene values.

Some of the important feature selection algorithms which are currently being used in the field of the microarray data analysis [39] are the following: new robust feature selection method, entropic filtering algorithm (EFA), MASSIVE, maximum weight and minimum redundancy (MWMR), minimum redundancy maximum relevance (mRMR), INTERACT, Information Gain, ReliefF, Correlation-based Feature Selection, Fast Correlation-Based Filter, new hybrid filter-based FS based on the combination of clustering and modified Binary Ant System (FSCBAS), particle swarm optimization, ant colony optimization, genetic algorithm, artificial bee colony, unsupervised feature selection approach based on ACO (UFACO), relevance–redundancy FS using ACO, a binary ant colony optimization (BACO), novel hybrid feature selection called R-m-GA, Multi-Filter Multi-Wrapper (MFMR), SVM-RFE, iterative perturbation method (IFP), First Order Inductive Learner, rule-based feature subset selection algorithm, kernel penalized SVM (KP-SVM), Adaptive Genetic Algorithm, and Mutual Information Maximization (MIM).

3.6   Importance of Feature Selection in Streaming Data Analysis

Streaming feature selection is an emerging research area which is being currently focused on by researchers to reduce the features and select the most informative features. In streaming feature selection, the candidate features arrive in a sequential manner and also the size of the features will be unknown. This type of streaming feature selection has been used in many application areas such as weather forecasting, stock market prediction, and clinical record analysis. The streaming features are defined as features which flow one by one based on the time variations, but the number of instances is fixed. Because features flow one by one, decisions have to be made whether the feature has to be kept or discarded. Noura AlNuaim et al. [40] enumerated the differentiation between the streaming data and the streaming features. In streaming data, the number of features is fixed and the instances of the streaming data will be generated automatically over time, and hence, the size of instances is unknown. In streaming features, the number of instances is fixed where the number of features will be changed over the time.

Peng Zhou et al. [41] proposed a novel neighborhood rough set-based feature selection with adapted neighbors called gap relation and new online feature selection method called OFS-A3M. The main novelty of this paper is that the proposed OFS-A3M does not require any of the domain knowledge in advance. There are three evaluation criteria, namely maximal-relevance, maximal-dependency, and maximal significance that were considered and used to select the optimal features

Table 8 Feature selection algorithms on video processing

Ref. no.	Dataset	Feature selection algorithm	Classifiers used	Accuracy	Year
[42]	LUNG2, IONOSPHERE, ARCENE WDBC, SRBCT, LYMPHOMA, SONAR, HILL, COLON, GLIOMA, MLL, PROSTATE, DLBCL, LEU,	OFS-Density based on neighborhood rough set	KNN, SVM, and CART	Average KNN – 85.56% SVM – 84.18% CART – 80.80%	2019
[43]	ALLAML, GLIOMA, Prostate GE, Breast, SRBCT, CNS + 20 datasets	OSFSMI and OSFSMI-k	Naïve Bayes KNN Decision Tree	Naïve Bayes – 58.64% KNN – 68.97% Decision Tree – 64.09% Average	2018
[39]	Dorothea, arcene, dexter, and madelon, nova, sylva, and hiva, arrhythmia and mf, tm1, tm2, and tm3	OS-NRRSAR-SA	J48, SVM, Naive Bayes	Naive Bayes – 46.03– 98.90% SVM – 67.32– 67.32% J48 – 51.72– 98.20%	2016

and also, it selects the predominant features based on high dependency, high correlation, and low redundancy. In their studies, fifteen different types of datasets were used to compare the results of the proposed algorithm. Some the standard streaming feature selection algorithms are as follows: information-investing and alpha-investing based on streamwise regression for online feature selection, online streaming feature selection framework with two algorithms called fast Online Streaming Feature Selection (OSFS) and OSFS, streamwise feature selection from the Rough Set perspective, OS-NRRSARASA- Rough Set-based method for online streaming feature selection, OSFSMI and OSFSMI-k, novel Neighborhood Rough Set classifier (NRSC), Scalable and Accurate Online FS Approach (SAOLA), Grafting algorithm based on a stagewise gradient descent approach for online feature selection, Drift Detection Method and the Early Drift Detection Method, and ADaptive sliding WINdow (ADWIN).

Some of the recent video processing methods and applied datasets with accuracy are presented in Table 8.

4 Discussion

When concerning with the high-dimensional spaces of data, the learning phase suffers to observe or conceive the features since all the features do not have the same amount of discriminative power and consist of both redundant and irrelevant data. In this context, filter methods, wrapper methods, and embedded methods are optimally used to select the features with many learning algorithms and contribute in different domains. Filter-based methods extract features in preprocessing step so as to estimate each subset and properties of data. The main merit of this technique is that it is computationally fast and scalable to large-dimensional data and the disadvantage of this technique is that it has no connection with classifier to achieve better performance. The wrapper-based method uses the learning model as the black box to assess the features based on predictive ability. The advantage of this method is that it interacts with the feature subset as well as the learning model well, but the disadvantage is the existence of larger risk of overfitting problem. Embedded methods use both feature selection model and learning model for the purpose of classification, and also it produces good performance results. The advantage of this method is that it interacts with the feature subset as well as the learning model well, but the disadvantage is the existence of larger risk of overfitting problem. Embedded methods use both feature selection model and learning model for the purpose of classification, and also it produces good performance results.

As mentioned earlier, feature selection is used in various domains. In the medical domain, it plays a major role in predicting cancers in an early stage based on the epidemic changes of the genes by selecting the highly informative genes in a timely manner. In the perspective of analytics in video, it is used in predicting the crime, gait recognition, etc., and also in text mining, it is used in categorizing the texts in various domains and aspects. Feature selection plays an important role in streaming data selection for predicting stock marketing, weather forecasting, etc., and also used in the intrusion detection system for detecting intruders in the networks.

5 Conclusion

In this chapter, we have provided a detailed introduction to feature selection with state-of-the-art feature selection techniques based on filter, wrapper, embedded, and hybrid models. Moreover, we have provided the taxonomy of the dimensionality reduction techniques and fuzzy logic-based feature selection techniques in a detailed manner. Further, we have discussed the importance of feature selection among various application domains such as text analytics, video analytics, audio analytics, microarray data analysis, intrusion detection systems, and feature selection in stream data analysis.

References

1. Bolon-Canedo, V., Sanchez-Marono, N., Alonso-Betanzos, A., Benitez, J. M., & Herrera, F. (2014). A review of microarray datasets and applied feature selection methods. *Information Sciences, 282*, 111–135.
2. Wang, H., Tan, L., & Niu, B. (2019). Feature selection for classification of microarray gene expression cancers using bacterial colony optimization with multi-dimensional population. *Swarm and Evolutionary Computation, 48*, 172–181.
3. Bolón-Canedo, V., Sánchez-Maroño, N., Alonso-Betanzos, A., Benítez, J. M., & Herrera, F. (2019). A review of microarray datasets and applied feature selection methods. *Information Sciences, 282*, 111–135.
4. Ang, J. C., Mirzal, A., Haron, H., & Hamed, H. N. A. (2019). Supervised, unsupervised, and semi-supervised feature selection: A review on gene selection. *IEEE/ACM Transactions on Computational Biology and Bioinformatics, 13*, 971–989.
5. Nematzadeh, H., Enayatifar, R., Mahmud, M., & Akbari, E. (2019, January 17). Frequency based feature selection method using whale algorithm. *Genomics, 111*, 1946–1955.
6. González, J., Ortega, J., Damas, M., Martín-Smith, P., & Gan, J. Q. (2019). A new multi-objective wrapper method for feature selection–accuracy and stability analysis for BCI. *Neurocomputing, 333*, 407–418.
7. Kumar, L., & Bharti, K. K. (2019). An improved BPSO algorithm for feature selection. In *Recent trends in communication, computing, and electronics* (pp. 505–513). Singapore: Springer.
8. Cilia, N. D., De Stefano, C., Fontanella, F., & di Freca, A. S. (2019). A ranking-based feature selection approach for handwritten character recognition. *Pattern Recognition Letters, 121*, 77–86.
9. Rendall, R., Castillo, I., Schmidt, A., Chin, S. T., Chiang, L. H., & Reis, M. (2019). Wide spectrum feature selection (WiSe) for regression model building. *Computers & Chemical Engineering, 121*, 99–110.
10. Mafarja, M., Aljarah, I., Faris, H., Hammouri, A. I., Ala'M, A. Z., & Mirjalili, S. (2019). Binary grasshopper optimisation algorithm approaches for feature selection problems. *Expert Systems with Applications, 117*, 267–286.
11. Xiong, C. Z., Su, M., Jiang, Z., & Jiang, W. (2019). Prediction of hemodialysis timing based on LVW feature selection and ensemble learning. *Journal of Medical Systems, 43*(1), 18.
12. Singh, A., & Jain, A. (2019). Adaptive credit card fraud detection techniques based on feature selection method. In *Advances in computer communication and computational sciences* (pp. 167–178). Singapore: Springer.
13. Sayed, G. I., Hassanien, A. E., & Azar, A. T. (2019). Feature selection via a novel chaotic crow search algorithm. *Neural Computing and Applications, 31*(1), 171–188.
14. Anter, A. M., Azar, A. T., & Fouad, K. M. (2019, March). Intelligent hybrid approach for feature selection. In *International conference on Advanced Machine Learning Technologies and Applications* (pp. 71–79). Cham: Springer.
15. Chiew, K. L., Tan, C. L., Wong, K., Yong, K. S., & Tiong, W. K. (2019). A new hybrid ensemble feature selection framework for machine learning-based phishing detection system. *Information Sciences, 484*, 153–166.
16. Al-Tashi, Q., Kadir, S. J. A., Rais, H. M., Mirjalili, S., & Alhussian, H. (2019). Binary optimization using hybrid grey wolf optimization for feature selection. *IEEE Access, 7*, 39496–39508.
17. Zheng, Y., Li, Y., Wang, G., Chen, Y., Xu, Q., Fan, J., & Cui, X. (2019). A novel hybrid algorithm for feature selection based on whale optimization algorithm. *IEEE Access, 7*, 14908–14923.
18. Arora, S., Singh, H., Sharma, M., Sharma, S., & Anand, P. (2019). A new hybrid algorithm based on grey wolf optimization and crow search algorithm for unconstrained function optimization and feature selection. *IEEE Access, 7*, 26343–26361.

19. Mohan, C., & Nagarajan, S. (2019). An improved tree model based on ensemble feature selection for classification. *Turkish Journal of Electrical Engineering and Computer Sciences, 27*(2), 1290–1307.
20. Song, X., Waitman, L. R., Hu, Y., Yu, A. S., Robins, D., & Liu, M. (2019). Robust clinical marker identification for diabetic kidney disease with ensemble feature selection. *Journal of the American Medical Informatics Association, 26*(3), 242–253.
21. Bui, D. T., Tsangaratos, P., Ngo, P. T. T., Pham, T. D., & Pham, B. T. (2019). Flash flood susceptibility modeling using an optimized fuzzy rule based feature selection technique and tree based ensemble methods. *Science of the Total Environment, 668*, 1038–1054.
22. Fan, S., Tang, J., Tian, Q., & Wu, C. (2019). A robust fuzzy rule based integrative feature selection strategy for gene expression data in TCGA. *BMC Medical Genomics, 12*(1), 14.
23. Jiménez, F., Martínez, C., Marzano, E., Palma, J., Sánchez, G., & Sciavicco, G. (2019). Multiobjective evolutionary feature selection for fuzzy classification. *IEEE Transactions on Fuzzy Systems, 27*, 1085–1099.
24. Dzulkalnine, M. F., & Sallehuddin, R. (2019). Missing data imputation with fuzzy feature selection for diabetes dataset. *SN Applied Sciences, 1*(4), 362.
25. Arefnezhad, S., Samiee, S., Eichberger, A., & Nahvi, A. (2019). Driver drowsiness detection based on steering wheel data applying adaptive neuro-fuzzy feature selection. *Sensors, 19*(4), 943.
26. Guru, D. S., Suhil, M., Raju, L. N., & Kumar, N. V. (2018). An alternative framework for univariate filter based feature selection for text categorization. *Pattern Recognition Letters, 103*, 23–31.
27. Labani, M., Moradi, P., Ahmadizar, F., & Jalili, M. (2018). A novel multivariate filter method for feature selection in text classification. *Engineering Applications of Artificial Intelligence, 70*, 25–37.
28. Mannepalli, K., Sastry, P. N., & Suman, M. (2018). Emotion recognition in speech signals using optimization based multi-SVNN classifier. *Journal of King Saud University – Computer and Information Sciences.* https://doi.org/10.1016/j.jksuci.2018.11.012
29. Özseven, T. (2019). A novel feature selection method for speech emotion recognition. *Applied Acoustics, 146*, 320–326.
30. Srinivasa Murthy, Y. V., & Koolagudi, S. G. (2018). Classification of vocal and non-vocal segments in audio clips using genetic algorithm based feature selection (GAFS). *Expert Systems with Applications, 106*, 77–91.
31. Zhang, S., Zhang, S., & Huang, T. (2019). Speech emotion recognition using deep convolutional neural network and discriminant temporal pyramid matching. *IEEE Transactions on Multimedia, 20*(6), 1576–1590.
32. Shamim Hossaina, M., & Muhammad, G. (2019). Emotion recognition using deep learning approach from audio–visual emotional big data. *Information Fusion, 49*, 69–78.
33. Mao, Q., Dong, M., Huang, Z., & Zhan, Y. (2014). Learning salient features for speech emotion recognition using convolutional neural networks. *IEEE Transactions on Multimedia, 16*(8), 2203–2213.
34. Yan, Y., Shen, H., Liu, G., Ma, Z., Gao, C., & Sebe, N. (2014). GLocal tells you more: Coupling GLocal structural for feature selection with sparsity for image and video classification. *Computer Vision and Image Understanding, 124*, 99–109.
35. Bampis, C. G., & Bovik, A. C. (2018). Feature-based prediction of streaming video QoE: Distortions, stalling and memory. *Signal Processing: Image Communication, 68*, 218–228.
36. Zhou, H., You, M., Liu, L., & Zhuang, C. (2017). Sequential data feature selection for human motion recognition viaMarkov blanket. *Pattern Recognition Letters, 86*, 18–25.
37. Benuwaa, B.-B., Zhana, Y., Monney, A., Ghansah, B., & Ansah, E. K. (2019). Video semantic analysis based kernel locality-sensitive discriminative sparse representation. *Expert Systems with Applications, 119*, 429–440.
38. Selvakumar, K., Karuppiah, M., SaiRamesh, L., Islac, S. K. H., Hassan, M. M., Fortino, G., & Choo, K.-K. R. (2019). Intelligent temporal classification and fuzzy rough set-based feature selection algorithm for intrusion detection system in WSNs. *Information Sciences, 497*, 77–90.

39. Eskandari, S., & Javidi, M. M. (2016). Online streaming feature selection using rough sets. *International Journal of Approximate Reasoning, 69*, 35–57.
40. AlNuaimi, N., Masud, M. M., Serhani, M. A., & Zaki, N. (2019). Streaming feature selection algorithms for big data: A survey. *Applied Computing and Informatics*. https://doi.org/10.1016/j.aci.2019.01.001
41. Zhoua, P., Hua, X., Li, P., & Wu, X. (2019). Online streaming feature selection using adapted neighborhood rough set. *Information Sciences, 481*, 258–279.
42. Zhou, P., Hu, X., Li, P., & Wu, X. (2019). OFS-density: A novel online streaming feature selection method. *Pattern Recognition, 86*, 48–61.
43. Rahmaninia, M., & Moradi, P. (2019). OSFSMI: Online stream feature selection method based on mutual information. *Applied Soft Computing, 68*, 733–746.

Ontological Structure-Based Retrieval System for Tamil

S. Rajendran, K. P. Soman, Anandkumar M, and C. Sankaralingam

1   Introduction

Ontologies prove to be extremely useful in the representation of lexical knowledge. The renewed interest in lexical semantics and natural language processing (NLP) can be attributed to such researches. The meaning of a lexical item is partly determined by the position in the ontology occupied by the concept or concept it expresses. Representing one of the meanings of a word minimally implies (i) distinguishing it by other senses the same word might have, (ii) capturing certain inferences which can be performed from it and (iii) representing its similarity with the meaning of other words ([6], p. 31). Ontological representation is useful for this purpose. For instance, given the word *mouse,* a proper although minimal representation of its meaning requires distinguishing the sense of 'small rodent' from the one of 'small pointing device for computers'. Moreover, the same representation should be able to capture the fact that being a rodent entails being a mammal, as well as the fact that the sense of *mouse* as 'small rodent' shares with the meaning of other words such as *dog,* or *cat,* the fact of being subtypes of mammal. Ontologies are therefore powerful formal tools to represent lexical knowledge exactly. The word meanings can actually be regarded as entities to be classified in terms of the ontology types. In this perspective, a given sense can

S. Rajendran · K. P. Soman
Center for Computational Engineering & Networking (CEN), Amrita School of Engineering, Amrita Vishwa Vidyapeetham, Coimbatore, India

Anandkumar M (✉)
NITK, Surathkal, Mangalore, India
e-mail: m_anandkumar@nitk.edu.in

C. Sankaralingam
Tamil Virtual Academy, Chennai, Tamil Nadu, India

© Springer Nature Switzerland AG 2021   197
R. Kumar, S. Paiva (eds.), *Applications in Ubiquitous Computing*, EAI/Springer
Innovations in Communication and Computing,
https://doi.org/10.1007/978-3-030-35280-6_10

be described by assigning it to a particular type. The ontology structure will then account for entailments between senses in terms of relations between their types. The sharing of the same ontology type can be attributed to the resemblances between word senses ([6], p. 31).

2 Principles of Ontology

Hyponymy and its consequence taxonomy are the fundamental building blocks of ontology. Hyponymy is the relationship which exists between specific and general lexical items, such that the former is included in the latter. The set of terms which are hyponyms of same superordinate term (hypernym) are co-hyponyms. Take, for example, the lexical items *pacu* 'cow' and *vilanku* 'animal'. The propositions *atu oru pacu* 'that is a cow' and *atu oru erumai* 'that is a buffalo' unilaterally entail *atu oru vilanku* 'that is an animal'. The relationship existing between *pacu* 'cow' and *erumai* 'buffalo' with *vilanku* 'animal' is hyponymy and *pacu* 'cow' and *erumai* 'buffalo' are co-hyponyms and are the hyponyms of the hypernym *vilanku* 'animal'. The relation that holds between co-hyponyms is called incompatibility or opposition relation. Hyponymy and its natural partner, incompatibility, are described by Lyons [19] as the most fundamental paradigmatic relations of sense in terms of which the vocabulary is structured. Lyons [19] opines that taxonomic lexical hierarchies are structured by the relations of hyponymy and incompatibility. The relation of hyponymy imposes a hierarchical structure upon the vocabulary and upon particular fields within the vocabulary; and the hierarchical ordering of lexemes can be represented formally as a tree diagram ([19], p. 295). It is hard to conceive of any language operating satisfactorily in any culture without its vocabulary being structured in terms of the complementary principles of hyponymy and contrast ([19], p. 300).

Another important paradigmatic semantic relation holding lexical items is part-whole relation between which is known by the term meronymy. The relation has an inverse known by the term holonymy; if W_m is a meronym of W_h, then W_h is said to be a holonym of W_m. That is, if *hand* is a meronym of *body*, then *body* is the holonym of the meronym *hand*. The parts of the body such as *head*, *neck*, *chest*, *stomach*, *hands* and *legs* are the co-meronyms of *body*. For concrete objects like bodies and artefacts, meronymy can help to define them. There are numerous lexemes in the vocabularies of languages whose meaning cannot be specified independently of some part-whole relation of sense ([19], p. 314). Meronymic or partonomic relations are ontological relations that are considered as fundamental as the ubiquitous, taxonomic subsumption relationship ([29], p. 35).

Taxonomy is usually only a hierarchy of concepts (i.e. the only relation between the concepts is parent/child, or super-class/sub-class, or broader/narrower), but in an ontology, arbitrary complex relations between concepts can be expressed too (X married to Y; or A works for B; or C is located in D, etc.). Although taxonomy contributes to the semantics of a term in a vocabulary, ontologies include richer relationships between terms. It is these rich relationships that enable the expression of domain-specific knowledge, without the need to include domain-specific terms.

Ontology has a richer internal structure, as it includes relations and constraints between the concepts. The word *vocabulary* and *ontology* are often used interchangeably. But a more strict definition is that a vocabulary is a collection of terms being used in a particular domain that can be structured (e.g. hierarchically) as a taxonomy. This taxonomy, when combined with some relationships, constraints and rules, forms the ontology. A combination of ontology together with a set of instances of classes constitutes a knowledge base as given below ([5], p. 58):

Vocabulary + structure = Taxonomy
Taxonomy + Relationships, constraints and rules = Ontology
Ontology + instances = Knowledge base

2.1 Ontology of Aristotelian Origin

Aristotle instigated the history of scientific taxonomy. It is predicated on a first philosophy of essentialism. Aristotle's work on natural history and logic laid out taxonomic principles [42]. Taxonomy or the division of things into genera and species is a way of classifying predicates in the logic; it is a refinement of the ten basic categories of predicates. The basic idea was developed in detail in the long tradition of Aristotelian logic.

2.2 Ontology in Nida's Thesauri Dictionary

Nida [26] who was concerned with the preparation of a thesauri dictionary for Greek gives the following as the tentative hierarchical classification of the referential meanings or lexical concepts ([26], pp. 178–186). His is a componential approach to meaning. He elaborately discusses the foundation of his theory of classification in his work entitled 'Componential Analysis of Meaning: An Introduction to Semantic Structure' [26]. He has classified the lexical concepts under four categories: entities, events, abstracts and relationals.

I. Entities

  A. Inanimate

    1. Natural: (a) Geographical, (b) Natural substances, (c) Flora and plant products
    2. Manufactured or constructed entities: (a) Artefacts (non-constructions), (b) Processed substances: foods, medicines, and perfumes, (c) Constructions

    B. Animate entities

        1. Animals, birds, insects, 2. Human beings, 3. Supernatural power or beings

II. Events

    A. Physical, B. Physiological, C. Sensory, D. Emotive, E. Intellection, F. Communication, G. Association, H. Control, I. Movement, J. Impact, K. Transfer, L. Complex activities, involving a series of movements or actions

III. Abstracts

    A. Time, B. Distance, C. Volume, D. Velocity, E. Temperature, F. Colour, G. Number, H. Status, I. Religious character, J. Attractiveness, K. Age, L. Truth-falsehood, M. Good-bad, N. Capacity, O. State of health, etc.

IV. Relationals

    A. Spatial, B. Temporal, C. Deictic, D. Logical, etc.

This classification is based on referential meanings and it is not possible to obtain one-to-one correspondence between the semantic domain of classes and the grammatical classes. A paper thesaurus prepared by Rajendran [33, 34] and an electronic thesaurus prepared by Rajendran and Baskaran [38, 39] are based on the classification given by Nida [26].

2.3 Other Ontologies

Starting form Aristotelian ontological structure mentioned above, many oncological structures have been developed to meet the requirements of the concerned. There are many types of ontological structures proposed for different purposes; some of them are: ontology in thesaurus, ontology in *nikhandu* tradition, ontology in semantic fields [21], ontology in semantic networks [43], ontology in WordNet [35, 36] and ontology in generative lexicon. For want of space, they are not discussed here.

3 Literature Survey

A number of researches have been undertaken for the creation of ontology and ontological tools for Indian languages. A few of them which we find important are discussed here. Saraswathi et al. [41] proposed a bilingual information retrieval system for English and Tamil. Talita et al. [44] discussed the challenges and solutions for building domain ontologies for indigenous languages. Chaware and Rao [7] presented an integrated approach to ontological development methodology with a case study. Maheswari and Karpagam [20] proposed a conceptual framework for ontology-based information retrieval. Ananthapadmanaban and Srivatsa [1] created user profile ontology for Tamil Nadu tourism. Chaware and Rao [8] developed an ontology approach for cross-language information retrieval. Durga and Govardhan [11] developed an ontology-based text categorization for Telugu documents. Saraswathi et al. [41] designed a semi-automatic ontology tree which are created partially manual and completed dynamically. Bhatt and Bhattacharyya [3] proposed a domain-specific ontology extractor for Indian languages. Chaware and Rao [8] proposed an ontology-supported interference system for Hindi and Punjabi. Athira et al. [2] presents architecture of an ontology-based domain-specific natural language question answering system. Kaur and Sharma [16] presented a pre-processing of domain ontology graph generation system in Punjabi. Grover and Chawla [13] evaluated the ontology creation of tools in the Indian context. Ramakrishnan and Vijayan [40] made a study on the development of cognitive support features in recent ontology visualization tools. Thaker and Goel [45] proposed a domain-specific ontology-based query processing system for Urdu language. Jidge [14] developed a domain-specific ontology extractor for Marathi language. Rajendran and Anandkumar [37] developed a visual onto-thesaurus for Tamil based on lexical semantic principles.

Kaur and Sharma [17] made a review of the techniques of ontology and its usage in Indian languages. Jidge and Govilkar [15] made a survey on ontology-building methodologies and tools for Indian languages. Kaur and Sharma [17] conclude from their review that very little work has been done for developing ontology in Indian languages. They attribute the reason to the fact that the number of challenges for the construction of ontology for minority languages is many and varied. Another reason is the lack of knowledge about different Indian languages. Jidge and Govilkar [15] more or less share the opinion of Kaur and Sharma. They make a conclusion that either ontology can be built using CLIR technique or by k-partite graph for the minority languages. They advocate for the expansion of WordNet [25] for building ontologies.

In the light of the present scenario explained above, an attempt is made to build an ontological structure-based retrieval system for Tamil which is explained in this chapter.

4 Ontological Structure for Tamil Vocabulary

Following Nida [26], the vocabulary of Tamil [32] is grouped initially into four domains: entities which consist of referential meanings of concrete concepts, events which consist mainly of verbs, abstracts which consist mainly of adjectives and adverbs apart from abstract nouns and relationals which consist of functional words including postpositions, connectives and coordinators. These four domains are further hierarchically classified into sub-domains under which the lexical items are listed. One will be able to capture the meaning of the concerned lexical item from the domain to which it belongs in a hierarchical fashion. A full-fledged detail of building ontology for the vocabulary of a language is available in Nida [27].

4.1 Structuring of Vocabulary by Lexical Relations

Lexical semantics offers foundation for structuring vocabulary in terms of sense relations or lexical relations ([9], pp. 230–335; [10, 19]). In the NLP-oriented papers, the general practice is to avoid giving linguistic details based on which the system is built. But here, we would like to give the lexical semantics of building OST to make it more transparent.

4.1.1 Lexical Relations

There are at least four lexical or sense relations by which lexical items can be linked or related to one another in OST. They are synonymy, hyponymy, compatibility and incompatibility ([9, 19], pp. 84–111). A word acquires its referential meaning in being a member of a semantic domain by the common features it shares with other members in that domain, and having contrasting features which separate it from other members of the domain. It is the semantic relations among words, such as synonymy, hyponymy, compatibility and incompatibility, which help one to classify and organize words in terms of semantic domains in a structural fashion.

Synonymy: e.g. *puttakam* 'book': *nuul* 'book';
Hyponymy-hypernymy: e.g. *pacu* 'cow': *vilangku* 'animal';
Meronymy-holonymy: e.g. *uTal* 'body': *kaal* 'leg';
Compatibility: e.g. *cellappiraaNiv* 'pet': *ndaay* 'dog'
Incompatibility: *malai* 'hill': *maTu* 'water hole'; incompatibility leads to the relation called opposition which culminates into many types which are discussed below.

4.1.2 Lexical Inheritance

Hypernymy-hyponymy and meronymy-holonymy ensure a lexical item to inherit semantic features as exemplified below:

kuiyl 'quail' – *iniya kuraluTaiya kariya paRavai* 'a black bird with sweet voice'
paRavai – *irukaalkaLum alakukaLum uTaiya, uTalin iruppakkangkaLilum paRap-pataRku eeRRavakaiyil ciRakukaL koNTa vilangkinam* 'an animal with two legs and a beak and wings at its sides of its body for flying'
vilangkinam 'animal' – *taanaaka iyangkum uTal uRuppukaLum celluloos ilaata celkaLum uLLa – uyiringkaL'* living beings with parts functioning automatically and cell walls without cellulose'
uyirinam – *uyir vaazkiRa onRu* 'one which lives'

4.1.3 Lexical Oppositions

There are many types of oppositions ([9], pp. 197–263; [19], pp. 270–290; [10], pp. 165–176). They can be grouped into two types of oppositions or contrasts based on the number of items involved in the contrast: binary contrast or opposition and non-binary contrast or non-binary opposition. If the contrast is made between two lexical items, it is called binary contrast; if the contrast is made between more than two lexical items, it is called non-binary contrast. The binary-contrasts are listed as follows: gradable opposites, ungradable opposites, or non-gradable opposites, complementaries, privative opposites, equipollent opposites, converses, directional opposites, orthogonal opposites, antipodal opposites and non-binary oppositions. OST tries to cover all these kinds of lexical oppositions. Due to want of space, they are not discussed here (See [9, 19] for details).

4.1.4 Hierarchies

Hierarchies ([9], pp. 112–118) are of two types: branching hierarchies and non-branching hierarchies. The branching hierarchy (Fig. 1) shows tree structure, whereas non-branching hierarchy (Fig. 2) does not show tree structure.

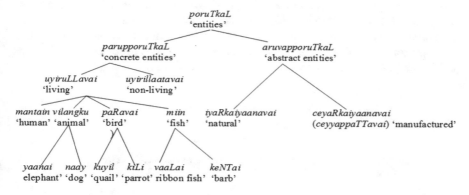

Fig. 1 Branching hierarchies

Fig. 2 Non-branching hierarchies

vaakkiyam 'sentence
|
eccattoTar 'clause'
|
toTar 'phrase'
|
col word'
|
urupan 'morpheme'

Taxonomic Hierarchies

The first major type of branching lexical hierarchy is the outcome of the hyponymy-hypernymy relation between lexical items. The sequence of hyponymy-hypernymy relation leads to branched-hierarchical structure of a set of vocabulary items which show this pair of relations among themselves. Taxonomic hierarchies (Fig. 3) ([9], pp. 136–155) are more liberal than hyponymy-hypernymy hierarchies. The following is an example.

Meronymic Hierarchies

The second major type of branching lexical hierarchy is the part-whole type which is called meronomies ([9], pp. 157–180). Meronymic hierarchies are the result of meronymy-holonymy relation shown by the lexical items. The meronymy-holonymy relation also gives hierarchical structure (Fig. 4) to a set of vocabulary. The following is an example.

Non-branching Hierarchies

Non-binary opposition leads to a number of types of non-branching hierarchies. They can be listed as: bipoles, bipolar chains, monopolar chains, degrees, stages, measures, ranks, sequences and cyclic sets or cycles. Also, there are organizations such as propositional series or grids and clusters. They need to be accommodated in OST (See [10] for details about non-branching hierarches).

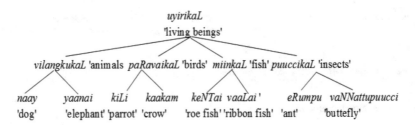

Fig. 3 Taxonomic hierarchy

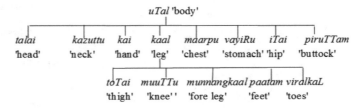

Fig. 4 Meronymic hierarchy

4.2 Creation of Database for OST

As mentioned earlier, the lexical items are arranged into four major categories entities, events, abstracts and relationals in line with Nida [27]. Each category requires different representation in the OST because of their inherent componential features. The organization of lexical items in the database is discussed below.

4.2.1 Organization of Entities in OST

Nida's [27] classification of entities is given above. Entities are represented as nouns in the surface level or formal level. It is proposed to make use of Nida's classification for organizing entities in OST. Relations pertaining to entities can be captured by lexical relations such as synonymy, hyponymy, compatibility, incompatibility and meronymy which have been elaborately discussed in the previous sections.

Entities in OST contain lexical items denoting concrete objects. Like any thesaurus, synonymy is captured in OST as a set of lexical items denoting the same meaning and labelled as *iNaiccoRkaL* 'the lexical items having the same meaning'. The notion of synonymy (referred here as *iNaimoziyam*) does not entail interchangeability in all contexts. By that criterion, natural languages have few synonyms. The more modest claim is that synonyms can be interchanged in some contexts. Although synonymy is a semantic relation between word forms, the semantic relation that is most important in organizing entities is the relation of subordination (or class inclusion or subsumption), which is called hyponymy. It is this semantic relation that organizes entities into a lexical hierarchy. Each hyponym leads on to a more generic hypernym. Hyponymy-hypernym relation cannot be represented as a simple relation between word forms. Hyponymy is a relation between lexicalized concepts, a relation that is represented in OST by the label *'uLLaTangkumoziym'* 'hyponymy' between the appropriate lexical concepts. A lexical hierarchy can be reconstructed by following the trail of hyponymically-hypernumically related lexical items.

Another important relation that helps in the tree representation of entities is part-whole relation. Part-whole relation between entries is generally considered to be a semantic relation, called meronymy. It is comparable to synonymy, antonymy and hyponymy. The relation has an inverse: if *x* is a meronym of *y*, then *y* is said to be a holonym of *x*. For concrete objects like bodies and artefacts, meronymy can help to define a basic level. Meronyms are distinguishing features that hyponyms can inherit. Consequently, meronymy and hyponymy are intertwined in complex ways. For example, if *alaku* 'beak' and *ciRaku* 'wing' are meronyms of *paRavai* 'bird', and if *kuyil* 'koel' is a hyponym of *bird*, then by inheritance, *beak* and *wing* must also be meronyms of *kuyil* 'koel'. In OST, the relation homonymy is referred as *uLLaTangkumoziyam* and the relation meronymy is referred as *pakutimoziyam*. The above-mentioned scenario (Fig. 5) is depicted in OST as follows:

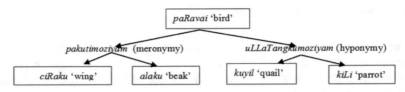

Fig. 5 Meronymy and hyponymy relation

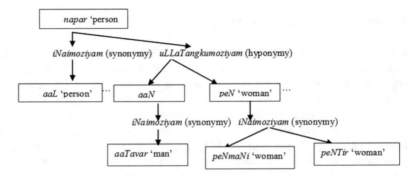

Fig. 6 Synonymy and hyponymy relation

The strongest psycholinguistic indication that two words are antonyms is that each is given on a word association test as the most common response to the other. Semantic opposition is not a fundamental organizing relation between nouns, but it does exist and so merits its own representation in OST. Take, for example, the following data structure: *ndapar* 'person', *aaL* 'person' {*aaN* 'male person', <*aaTavar* 'male person'>} *peN* 'female person', <*peNmaNi* 'female person', *peNTir* 'female person'>}. Note that hyponyms arranged (or included) under hypernym(s) by using the curly brackets ({}) and synonyms are arranged (or included) under the respective word-form by using the ankle brackets (< >). This gives the a structural representation (Fig. 6).

In Tamil, certain human entities have three forms: one is the epicene form, another is male form and the third is the female form; the epicene form implies a kind of respect when compared to the other two forms. For example, for the concept 'servant', there are three word forms in Tamil: *veelaikkaarar* 'servant' (epicene form), *veelaikkaaran* 'male servant' and *veelaikkaari* 'female servant'. In the tree structure (Fig. 7), the gender-marked forms are given under epicene form with the label *paalmoziyam* 'gender-marked relation' or 'gender-nymy'.

The relation among the three forms is established by giving the gender-marked forms under the epicene form by using the special separators.

Antonymy is a lexical relation between words, rather than a semantic relation between concepts. The antonyms are related to one another by using the special separators: % and *. For example, *peN* 'female person' is given under *aaN* 'female

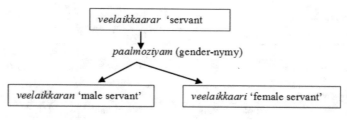

Fig. 7 Gender-nymy relation

Fig. 8 Antonymy relation

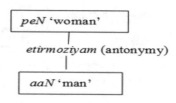

person' by using the above-mentioned separators as follows: *aaN %peN**. In the structure (Fig. 8), this is represented by using the label *etirmoziyam* 'antonymy' as given below.

When all three kinds of semantic relations – hyponymy, meronymy and antonymy – are included, the result is a highly interconnected network of entities.

The term 'function' has served many purposes, both in psychology and in linguistics. Nominal concepts can play various semantic roles as arguments of the verbs that they co-occur within a sentence [25]. For example, *katti* 'knife' – *veTTu* 'cut'; *kuzi* 'hole' – *tooNTu* 'dig'; *paTam* 'picture' – *varai* 'draw'. There are also linguistic reasons to assume that a thing's function is a feature of its meaning. It should be mentioned here that [30] in his generative lexicon talks about telic role in his qualia structure. At present, OST does not relate entities with their 'function' or telic role. This will be incorporated in OST later.

Table 1 gives the summary of the organization of entities in OST.

4.2.2 Organization of Events in OST

The semantic domain events comprise verbs and the abstract nouns derived from them. Nida's [28] tentative classification of events into twelve semantic domains based on componential analysis has been given already. Events are mostly realized in the surface level as verbal forms. Even though verbs do not show hierarchical ordering, a quasi-hierarchical ordering is possible by taking into account certain pertinent distinguishing semantic features. For wider coverage of verbs, it is proposed to follow the twelve-way classification of verbs by Nida [27] and this tentative classification is liable to change to accommodate more verbs.

Table 1 Summary of OST

Relations	Subtypes	Example
Synonymy		*puttakam* 'book' to *nduul* 'book'
Hypernymy-hyponymy		*vilangku* 'animal' to *paaluuTTi* 'mammal'
Hyponymy-hypernymy		*pacu* 'cow' to *paaluuTTi* 'mammal'
Holonymy-meronymy	Wholes to parts	*meecai* 'table' to *kaal* 'leg'
"	Groups to members	*tuRai* 'department' to *peeraaciriyar* 'professor'
Meronymy-holonymy	Parts to wholes	*cakkaram* 'wheel' to *vaNTi* 'cart'
"	Members to groups	*paTaittlaivar* 'captain' to *paTai* 'army'
Binary opposites	Antonymic (gradable)	*ndallavan* 'good person' to *keTTavan* 'bad person'
"	Complementary	*aaTavar* 'man': *makaLir* 'woman'
	Converse	*kaNavan* 'husband' *to manaivi* 'wife'
"	Privative (opposing features)	*ahRiNai* 'irrational' to *uyartiNai* 'rational'
"	Equipollent (positive features)	*aaN* 'male' to *peN* 'female'
"	Reciprocal social roles	*vaittiyar* 'doctor' to *ndooyaaLi* 'patient'
"	Kinship relations	*ammaa* 'mother' to *makaL* 'daughter'
"	Antipodal opposition	*cikaram* 'peak' to *aTi* 'foot (of mountain)'
	Orthogonal opposition	*ciRumi* 'girl': *ciRuvan* 'boy' *and peNTir* 'woman'
"	Degrees	*paaRai* 'mound': *kunRu* 'hillock': *malai* 'hill': *maamalai* 'mountain'
"	Ranks	*virivuraiyaaLar* 'lecture', *mutunilai virivuraiyaaLar* 'senior lecture', *iNaippeeraaciriya* 'reader', *peeraaciriyar* 'professor'
Compatibility		*ndaay* 'dog' to *cellappiraaNi* 'pet'

Polysemous Nature of Verbs

The verbs are fewer in number than nouns in Tamil and at the same time verbs are more polysemous in nature than nouns [31]. The semantic flexibility of verbs makes the lexical analysis of verbs difficult. A look at the Tamil corpus or Tamil dictionary will reveal the polysemous behaviour of verbs. Nida elaborately discusses the representation of polysemy of the verb *run* ([27], pp. 138–150) in his thesaurus. The polysemy will be captured in line with Nida [26] in OST.

Componential Features of Verbs

Verbs can be paraphrased in terms of finer semantic features. The decompositional nature of verbs can be exploited for the interpretation of verbs denoting complex events in terms of verbs denoting simple events. For example, the verb *kol* 'kill' can be decomposed into 'cause not to become alive'. The verb *eRi* 'throw' can be decomposed as 'cause an object to move away from one's possession by force'. The decompositional nature of verbs reveals the entailment relation existing between verbs. For example, the entailment of simple verb under causative verb (e.g. *ooTu* 'run' vs *ooTTu* 'cause to run') is understood by decompositional nature of verbs. The decompositional features of verbs can be captured by the componential analysis of verbs into finer semantic components [18]. All types of lexical relations such as synonymy, entailment, hyponymy and troponymy and sentential properties such as presupposition, inconsistency, tautology, contradiction and semantic anomaly can be mapped clearly if verbs are decomposed into componential features. The decompositional means of relating verbs or events will be considered in OST.

Synonymy Among Verbs

Synonymy is a rare phenomenon in verbal domain. Verbal domain exhibits only a few truly synonymous verbs. Take, for example, the words *paTi* 'read' and *vaaci* 'read'. *avan puttakam paTikkiRaan* 'He is reading a book' can entail *avan puttakam vaacikkiRaan* 'He is reading a book'. The relation existing between *paTi* and *vaaci* is synonymy and *paTi* and *vaaci* are synonyms, at least in this context. Truly synonymous verbs are difficult to find, and mostly quasi synonymous verbs are found in Tamil. The existence of a simple and a parallel compound forms (noun + verbalizer) prompts synonymy (quasi synonymy) in verbal system of Tamil. For example, *kol* 'kill' and *kolai cey* 'murder', *vicaari* 'enquire' and *vicaaraNai cey* 'investigate'. OST makes use of synonymy to relate one verb (verbal concept) with another verb (verbal concept) whenever it is possible.

Lexical Entailment and Meronymy

Lexical entailment refers to the relation that holds between two verbs by the statement 'X entails Y'. For example, *kuRaTTai viTu* 'snore' lexically entails *tuungku* 'sleep' because the sentence *avan kuRaTTai viTukiRaan* 'he is snoring' entails *avan tuungkukiRaan* 'he is sleeping'; the second sentence is true if the first one is true. Lexical entailment is a unilateral relation: if a verb V1 entails another verb V2, then it cannot be that case that V2 entails V1. For example, *uRangku* need not entail *kanavukaaN*.

Componential analyses have shown that verbs cannot be broken into referents denoted solely by verbs. It is true that some activities can be broken down into sequentially ordered sub-activities, say, for example, *camai* 'cook' is a complex activity involving a number of sub-activities. Consider the relation between the verbs *vaangku* 'buy' and *koTu* 'pay'. Although neither activity is a discrete part of the other, the two are connected, in that when you buy something, somebody gives it to you. Neither activity can be considered as a sub-activity of the other. Consider the relations among the activities denoted by the verbs *kuRaTTaiviTu* 'snore', *kanavukaaN* 'dream', and *uRangku* 'sleep'. Snoring or dreaming can be part of sleeping, in the sense that the two activities are, at least, partially, temporally co-extensive; the time that you spend snoring or dreaming is a proper part of the time you spend sleeping. And it is true that when you stop sleeping, you also necessarily stop snoring or dreaming. The relation between pairs like *vangku* 'buy' and *koTu* 'pay' and *kuRaTTaiviTu* 'snore' and *uRagnku* 'sleep' is due to the temporal relations between the members of each pair. The activities can be simultaneous (as in the case of *vaangku* 'buy' and *koTu* 'pay') or one can include the other (as in the case of *kuRaTTaiviTu* 'snore' and *uRangku* 'sleep'). OST makes of entailment to relate one verb with another verb whenever it is possible.

Hyponymy Among Verbs

Some verbs seem more generic than others. For example, *koTu* 'give' describes a wider range of activities than *viniyooki* 'distribute'. The hyponymous relation of the kind found in nouns cannot be realized in verbs. The sentence frame, *An x is a y*, which is used to establish hyponymous relation between nouns is not suitable for verbs, because it requires that x and y be nouns. The scrutiny of hyponyms and their superordinates reveals that lexicalization involves different kinds of semantic expansions across different semantic domains [12, 25]. Miller [22, 25] makes use of the term troponymy to establish this type of relation existing between verbs: 'When two verbs can be substituted into the sentence frame To V1 is to V2 in a certain manner, then V1 is a troponym of V2' ([22], p. 228). For example, *ndoNTu* 'to walk unevenly' is a troponym of *ndaTa* 'walk' as the former entails the latter.

Troponymy and Entailment

Troponymy is a particular kind of entailment in that every troponym of a more general verb X also entails Y [12, 25]. Consider, for example, the pair *ndoNTu* 'limp' and *ndaTa* 'walk'. The verbs in this pair are related by troponymy: *ndoNTu* is also *ndaTa* in a certain manner. So *ndoNTu* is a troponym of *ndaTa*. The verbs are also in entailment relation: the statement *avan ndoNTukiRaan* 'he is limping' entails *avan ndaTakkiRaan* 'he is walking'.

In contrast with pairs like *ndoNTu* 'limp' *and ndaTa* 'walk', a verb like *kuRaTTaiviTu* 'snore' entails and is included in *tuungku* 'sleep', but is not a troponym of *tuungku*. Similarly, *vaangku* 'buy' entails *koTu* 'give', but is not a troponym of *koTu* 'give'. The verbs in the pairs like *kuRaTTaiviTu* 'snore' and *tuungku* 'sleep' are related only by entailment and proper temporal inclusion.

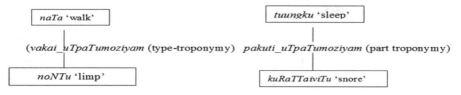

Fig. 9 Troponymy relation

It can be generalized that the verbs related by entailment and proper temporal inclusion cannot be related by troponymy. If the activities denoted by two verbs are temporally co-extensive, they can be linked by troponymy. Troponymy represents a special kind of entailment. In OST, the lexical items linked by troponymy are plotted in tree structure as given below. The relation between *ndaTa* amd *ndoNTu* is referred as *vakai_uTpaTumoziyam* 'type-troponymy' and the relation between *tungku* and *kuRaTTaiviTu* are referred as *pakuti_uTpaTumoziyam* 'part-troponymy'. The diagram (Fig. 9) depicts these relations.

OST makes use of paraphrases (or descriptions) to relate certain pairs of verbal concepts or events.

Opposition Relations and Entailment

Opposition relations are psychologically significant not only for adjectives, but also for verbs. It is found that after synonymy and troponymy, opposition relations are the most frequently coded semantic relations in building database for verbs [12, 25]. The semantics of opposition relations among verbs is complex. As for as Tamil is concerned, there is no morphologically derived opposite verbs. Some of the oppositions found among nouns are absent in verbs. A number of binary oppositions have been shown by the verbs that include converseness, directional, orthogonal, and antipodal oppositions. Active and passive forms of transitive verbs can be taken as showing converse opposition. *avan avaLaik konRaan* is in converse relation with the passive expression *avaL avanaal kollappaTTaaL*. Thus, active-passive pairs of transitive verbs in Tamil show converse opposition. The relation between the verbs *vaangku* 'buy' and *vil* 'sell' is rather more complex. The lexical items that are directionally opposite are in directional opposition. The relationship which hold between the pairs such as *vandtuceer* 'arrive' and *puRappaTu* 'reach', *vaa* 'come': *poo* 'go' is directional opposition. Under this category are the verb pairs such as *uyar* 'rise' and *taaz* 'go down', *eeRu* 'ascend' and *iRangku* 'descend'. There are other oppositions with reference to change of state, manner, speed, etc. as shown by the following pairs of examples: *kaTTu* 'build': *iTi* 'demolish', *kaTTu* 'tie': *aviz* 'untie', *ottukkoL* 'agree': *maRu* 'disagree', *uLLizu* 'inhale': *veLiviTu* 'exhale', *ndaTa*: *ooTu* 'run'. The diagram (Fig. 10) shows the opposition between verbs as represented in OST:

Fig. 10 Antonymy relation

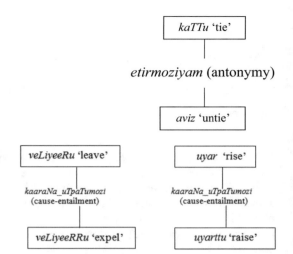

Fig. 11 Cause entailment

Not only the opposing features, even the presence or absence of a feature can also keep two items in opposition relation. These contrasting or distinguishing features can be arrived at by componential analysis of verbs [31]. The componential analysis of verbs shows that many verb pairs in an opposition relation also share an entailed verb. For example, the pair *jeyi/vel* 'succeed' and *tool* 'fail' entails *muyal* 'try'. The relation between the fist and the second are referred as *muRkooL_uTpaTumoziyam* 'presupposed-troponymy'.

Causation and Entailment

The causative relation exists between two verbal concepts: one is causative (e.g. *koTu* 'give') and the other is resultative (e.g. *peRu* 'get'). Causation can be considered as a specific kind of entailment as denoted by the following examples [12, 25]. *veLiyeeRRu* 'expel' entails *veLiyeeRu* 'leave', *uyarttu* 'raise' entails *uyar* 'rise' (temporal inclusion).

We have distinguished four different kinds of lexical entailment that systematically interact with the semantic relations mapped in OST. The concerned pairs are linked by the relation referred as *kaaraNa uTpaTumozi* 'cause entailment'. The following diagram (Fig. 11) shows the cause entailment relation.

Syntactic Properties and Semantic Relations

In recent years, there is a trend of incorporating syntactic properties in the lexicon itself. Viewing verbs in terms of semantic relations can also provide clues to an understanding of the syntactic behaviour of verbs. Incorporating the syntactic

Table 2 Summary of lexical relations

Relations	Definition/sub types	Example
Synonymy	Replaceable events	*tuungku* 'sleep' → *uRangku* 'sleep'
Meronymy-hypernymy	From events to superordinate events	*paRa* 'fly' → *pirayaaNi* 'travel'
Troponymy	From events to their subtypes	*naTa* → *noNTu* 'limp'
Entailment	From events to the events they entail	*kuRaTTaiviTu* 'snore' muyal 'try' *tuungku* 'sleep'
"	From event to its cause	*uyar* 'rise' → *uyarttu* 'raise'
"	From event to its presupposed event	*vel* 'succeed' → *muyal* 'try'
"	From even to implied event	*kol* 'murder' → *iRa* 'die'
Antonym	Opposites	*kuuTu* 'increase' → *kuRai* 'decrease'; *kaTTu* 'build' *iTi* 'demolish'
"	Conversensess	*vil* 'sell' → *vaangku* 'buy'
"	Directional opposites	*puRappaTu* 'start' → *vandtuceer* 'reach'
Derivatives	Verb to verbal noun	*paTi* 'study' → *paTippu* 'education'

properties of verbs in OST has to be explored for the better understanding of verbal concepts or events. Due to want of space, an elaborate discussion is avoided here.

Summing up of Relations of Events in OST

Table 2 sums up the lexical relations to be captured in the verb net.

4.2.3 Organization of Abstracts in OST

Nida considers abstracts as meanings which can be realized at the outset as adjectives and adverbs. Abstracts in OST contain mainly adjectives and adverbs apart from abstract nouns. Noun modification is primarily associated with the syntactic category 'adjective'. Similarly, verb modification is associated with the syntactic category 'adverbs'. Adjectives have their sole function of modification of nouns, whereas modification is not the primary function of noun, verb and prepositional phrases. The lexical organization of adjectives is unique to them, and differs from that of the other major syntactic categories, noun and verb. Three types of adjectives can be distinguished: Descriptive adjectives (Ex. *periya* big, *kanamaana* 'heavy'), Relational adjectives (Ex. *poruLaataara* 'economic', *cakootara* 'fraternal'), Reference-modifying adjectives (Ex. *pazaiya* 'old', *munndaaL* 'former') [24].

Adjectives

Descriptive Adjectives

A descriptive adjective is one that ascribes a value of an attribute to a noun. For example, *atu kanamaana cumai* 'that luggage is heavy' presupposes that there is attribute *eTai* 'WEIGHT' such that *eTai* (*cuma*i 'luggage') = *kanam* 'heavy'. In the same way, *taazndta* 'low' and *uyarndta* 'high' are values of HEIGHT [24]. Relating descriptive adjectives with the particular noun they pertain to is known by the term pertainymy. OST has to link the descriptive adjectives with the appropriate attributes.

Antonymy in Adjectives

Antonymy is the basic semantic relation that exists among descriptive adjectives. The word association tests reveal the importance of antonymy in adjectives ([23, 24], pp. 48–52). As the function of descriptive adjectives is to express values of attributes, and that nearly all attributes are bipolar, antonymy becomes important in the organization of descriptive adjectives. Antonymous adjectives express opposing values of an attribute. For example, the antonym of *kanamaana* 'heavy' is *ileecaana* 'light' that expresses a value at the opposite pole of the *kanam* 'WEIGHT' attribute [23, 24]. Antonymy, like synonymy, is a semantic relation between word forms. The problem is that the antonymy relation between word forms as illustrated in (Fig. 12) is not the same as the conceptual opposition between word senses.

Similarity in Adjective

Adjectives show bipolar structure ([23, 24], pp. 50–52). A set of adjectives show similarity of meaning with an adjective which is antonymous with another set of adjectives which show similarity of meaning with another adjective. The following examples show the existence of a bunch of adjectives denoting hotness against the bunch of adjectives denoting coldness. We can link them through their typical representatives as shown below. Similarity relation is mentioned in OST as *ottamoziyam* 'similarity-nymy'. The relation *ottamoziyam* 'similarity-nymy' is different form *iNaimoziyma* 'synonymy' as illustrated in (Fig. 13).

Fig. 12 Derived-nymy relation

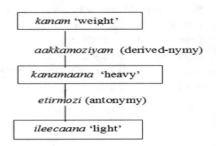

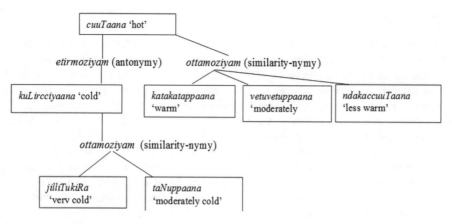

Fig. 13 Similarity-nymy relation

Table 3 Lexical memory for adjectives

cuuTu 'warmth'	vayatu 'age'
kotikkiRa 'very hot'	vayataana 'old'
cuuTaana 'hot'	ndaTuttara vayataana 'middle aged'
vetuvetuppaana 'warm'	iLamaiyaana 'young'
iLanjcuuTaana 'warm'	
kuLirndta 'cold'	

It can be inferred that *periya* 'big' and *ciRiya* 'small' form an antonymous pair and *ndiiNTa* 'long' and *kuTTaiyana* 'short' form another antonymous pair. These pairs demonstrate that antonymy is a semantic relation between words rather than concepts.

Gradation in Adjectives

Gradation ([23, 24], pp. 52–53) is one of the important properties found among adjectives. Most discussions on antonymy distinguish between contradictory and contrary terms. This terminology originated in logic, where two propositions are said to be contradictory if the truth of one implies the falsity of the other and are said to be contrary if only one proposition can be true but both can be false. For example, *uyiruLLa* 'alive': *cetta* 'dead' (contradictory terms) and *kuNTaana* 'fat': *melindta* 'thin' (contrary terms). Contraries are gradable adjectives, contradictories are not. Gradation therefore must also be considered as a semantic relation organizing lexical memory for adjectives (Table 3).

For some, attributes of gradation can be expressed by ordered strings of adjectives, all of which point to the same attribute noun in OST.

Markedness in Adjectives

Markedness ([23, 24], pp. 53–54) is an important property found among adjective. Binary oppositions frequently have a marked term and an unmarked term. That is, the terms are not entirely of equivalent weights, but one (the unmarked one) is neutral or positive in contrast to the other. Marked vs unmarked distinction is found in polar oppositions such as the following: *uyarndta* 'high': *taazndta* 'low', *vayataana* 'old': *iLamaiyaana* 'young', *niiLamaana* 'long': *kuTTaiyaana* 'short', *akalamaana* 'wide': *kuRukalaana* 'narrow'. We talk about the concept of 'height' in terms of *uyaram* 'height' rather than *kuTTai* 'short'. So, the primary member *uyaramaana* 'high' is considered as the unmarked term; the secondary member, *kuTTaiyaana* 'short' is the marked one. They are related to the attribute noun *uyaram* 'height'. OST captures the relation between marked and unmarked terms and their cross-reference to their variable property as shown in (Fig. 14).

Other Aspects of Adjectives

The colour adjectives, reference-modifying and referent-modifying adjectives and relational adjectives including polysemy in adjectives need to be dealt differently in OST. Due to want of space, they are not discussed here (See [4, 23, 24] for further details).

Adverbs

Most of the adverbs are derived from nouns by adding suffix in the similar way adjectives are derived from nouns. The derived adjectives and adverbs need to be linked with the nouns from which they are derived. For example, *aazak-aana* 'beautiful' and *azak-aaka* are derived from the noun *azaku* 'beauty'; similarly, *veekam-aana* 'fast (adj.)' and *veekam-aaka* 'fast (adv.)' are derived from the noun *veekam* 'speed'. This is captured in OST by linking the derived forms with the noun by the relation *aakkamoziyam* 'derivative relation' or 'derived-nymy' as shown in the following diagrams (Fig. 15).

The derived adjectives and adverbs inherit the semantic property of the noun from which they are derived. The semantic organization of adverbs is simple and straightforward. There is no tree structure ([23, 24], p. 61).

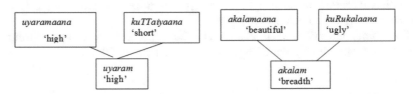

Fig. 14 Inheriting semantic property

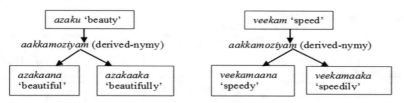

Fig. 15 Derived-nymy

Table 4 Summary of relations in abstracts

Relations	Subtypes	Example
Synonymy		*tukkam* 'sorrow' to *tunpam* 'sorrow', *cangkaTam* 'sorrow', *tuyaram* 'sorrow'
Hypernymy-hyponymy		*uNarcci* 'feeling' to *makizcci* 'happiness'
Hyponymy-hypernymy		*paccai* 'green' to *niRam* 'colour'
Holonymy-meronymy	Wholes to parts	*vaaram* 'week' to *ndaaL* 'day'
Meronymy-holonymy	Parts to wholes	*injcu* 'inch' to *aTi* 'feet'
Binary opposites	Antonymic (gradable)	*ndalla* 'good' to *keTTa* 'bad'
"	Temporal relations	*munnar* 'before' to *pinnar* 'after'
"	Orthogonal or perpendicular opposition	*vaTakku* 'north' to *kizakku* 'east' and *meeRku* 'west'
"	Antipodal opposition	*vaTakku* 'north' to *teRku* 'south'
Multiple opposites	Serial	*onRu* 'one', *iraNTu* 'two', *muunRu* 'three', *ndaanku* 'four'
"	Cycle	*njaayiRu* 'Sunday' to *tingkaL* 'Monday' ... to *cani* 'Saturday'

Relations	POS linked	Example
Antonymy (gradable, i.e. contrary)	Adjective-adjective	*azakaana* 'beautiful': *kuruurmaana* 'ugly'
Antonymy (non-gradable, i.e. contradictory)	Adjective-adjective	*uyiruLLa* 'alive': *cetta* 'dead'
Derivational	Adjective-noun	*azakaana* 'beautiful': *azaku* 'beauty'
Attributive	Noun-adjective	*vaTivam* 'size': *cinna* 'small'
Relational	Adjective-noun	*poruLaataara* 'economical': *poruLaataaram* 'economy'
Similarity	Adjective-adjective	*paaramaana* 'heavy': *kanamaana* 'heavy'
Derivational	Noun-adjective/adverb	*azaku* 'beauty': *azakaana* 'beautifull', *azakaaka* 'beautifully'
Similarity	Adverb-adverb	*veekamaaka* 'fast': *viraivaaka* 'fast'

Summing up of Relations in Abstracts

Table 4 summarizes the relations in abstracts.

Table 5 Compilation of OST

Meaning relations	Their Tamil equivalents	Relation distinguishing symbols used in the database
Synonymy	இணைமொழியம்/iNaimoziyam	<>
Hyponymy	வகைமொழியம்/vakaimoziyam	{ }
Meronym	பகுதிமொழியம்/pakutimoziyam	/ |
The relation between the epicene noun and the gender-marked nouns.	பால்மொழியம்/paalmoziyam	$#
Antonymy	எதிர்மொழியம்/etirmoziyam	*
The relation between the verb and the nouns derived from the verb.	பெயராக்கமொழியம்/peyaraakka moziyam	~ '
The relation between a noun and the modifiers (adjectives and adverbs) derived from the noun	அடைமொழியம்/aTaimoziyam	^ !

4.3   Compilation of OST

So far, we have explained about the OST in a wider perspective. But the present OST built by us has limited scope. Due to want of fund and other resources, we have not fully achieved the building OST as explained above. We have not covered all the semantic or lexical relations between words as explained above. Table 5 gives the description about the compilation of OST with limited scope.

4.3.1   Data Base Structure of OST

The table 5 gives the different types of meaning relations established between words or lexical items and the unique way of representing them in the database structure for the ontology tree generation.

Sample of Data Base Structure for Synonymy

The synonyms are given inside the angle brackets (<>) in the database structure of OST. The following is a sample depicting the synonymy of *cuuriyan* 'sun' with its co-synonyms:

cuuriyan   <njaayiRu, aatavan, pakalavan, tivaakaran, aatittan, katiravan, katiroon, kiraNan, cengkatiroon, cengkatir, vengkatiroon, vengkatir, venjcuTar, aayirangkatiroon, aayirangkiraNan, ulakaneentiran, ulakappaantavan>; {ezunjaayiRu, eeRunjaayiRu; caaynjaayiRu, iRangku njaayiRu; iLanjcuuriyan, <paalacuuriyan>;}

Sample of Data Structure for Hyponymy-Hypernymy

The hyponyms of a hypernym are given inside the curly brackets ({ }) in the database structure of OST. The following is the sample of the database structure for the hypernym *ndvakirakam* 'nine-planet' with its hyponyms:

ndavakirakam {cuuriya; candtiran; putan; viyaazan; cukkiran; cani}

Sample of Data Stricter for Meronymy-Holonymy

The meronyms of a holonym are given inside the pair of symbols – / and | – in the database structure of OST. The following is the sample of database structure of the holonym *uTal* 'body' with its meronyms:

uTal/talai; kazuttu; maarpu; vayiRu; piruttam; kai; kaal|

Sample of Data Structure for Antonymy

The antonyms are related to one another in the database of the OST by linking them using the pair of symbols – % and *. The following is the sample of database structure depicting the antonymous relation of the following pairs of concepts: 'dark vs bright', 'short vs long' and 'hot vs cold'.

iruTTaana %veLiccamaana*; kuTTiyaana %niiLamaana*; cuuTaana %kuLirc-ciyaana*

Sample of Data Structure for Gender-nymy

In Tamil, there are many nouns which can be said as the epicene form (non-gender-marked forms) of the gender-marked nouns. For example, *veelaikaarar* 'servant' has the male gender-marked form *veelaikkaaran* 'male servant' and female gender-marked form *veelaikkaari* 'servant woman'. The relation between the gender-marked ones and the non-gender-marked one is called gender-nymy. The gender-marked nouns are related to their corresponding non-gender-marked epicene noun using the pair of symbols – $ and #. The following is the sample of database structure depicting gender-nymy of the concepts 'male young person' and 'female young person' with the non-gender-marked concept 'young person'.

iLainjarkaL: {iLainjar $iLainjan#, vaalipar $vaalipan#, vaalippaiyan, iLaval, iLantaari, iLainjoor $iLainjoon# kaaLai, viTalai, iLamvaTTam, vayacuppaiyan, vayacuppiLLai, vaycuppiLLaiyaaNTan, iLamvayatinar $iLamvayatinan# iLampiraayattaar $ilampiraayattaan#, patinparuvattinar $patinparuvattinan#, *patinvayatinar $patinvayatinan#, vayatuvantoor, vayatuvantavar $vayatuvanta-van#, iLantai, iLantaari, vaaliyan, paaliyan; iLarattam;>}*

Sample of Data Structure for Derived-nymy

There many nouns (i.e. verbal nouns) which can be considered as the derivatives of their respective verbs. The relation of the verbal nouns with the verb is called derived-nymy. The verbal nouns are linked to their parent verb by using the pair of symbols – ~ and '. Following is the sample of database structure of the derived-nymy of the verbal nouns *ndikaztal* 'happening' and *ndikazcci* 'event' with the verb *ndikaz* 'happen'.

ndikazvukaLkuRittavan: {ndikaz; ~ndikaztal; ndikazci; ndikazvu; '<campavi; ~campavittal; campavam; ' viLai;> ~viLaital; viLaivu;' ndeer; ~ndeertal;' palittal; ~palittal;' iiTeeRu; ~iiTeeRutal; iiTeeRRam;' ndilavu; ~ndilavutal;'}

Sample of Data Structure for Attributive-nymy

There are many adjectives and adverbs which can be considered as derived from their source nouns. For example, the adverb *azakaaka* 'beautifully' and the adjective *azakaana* 'beautiful' are derived from the noun *azaku* 'beauty'. The adjectives and adverbs derived from the concerned nouns are given inside the pair of symbols – ˆ and !. The following is the sample of base structure of attributive-nymy of adjectives and adverbs with noun *ndiiNTakaalam* 'long period of time'.

ndiiNTakaalamtoTarpaanavai: {ndiiNTakaalam; ˆndiiNTakaalamaaka!; ndiiTuuzi, <ndeTungkaalam ˆndeTungkaalamaaka;! rempakaalam; ˆrempakaalamaaka,! niiNTaneeram, <ndeTuneeram, rempaneeram, niRaiyaneeram;> ndiiN-TakaaL, <neTunaaL; rempanaaL, ndiRaiyanaaL>; pallaaNTu; nduuRRaaNTu, <nduRaaNTu>}

4.3.2 User Interface of OST

Generating ontology tree from a given huge database structure for a user's query is the backbone activity for the OST which is an intelligent information retrieval and visualization system. This type of tool provides a powerful way for data representation and knowledge mining process. The user can input search word to the system and extract the relevant information from the database (Fig. 16). The information is given in a visual representation. A well-structured database is needed for this purpose.

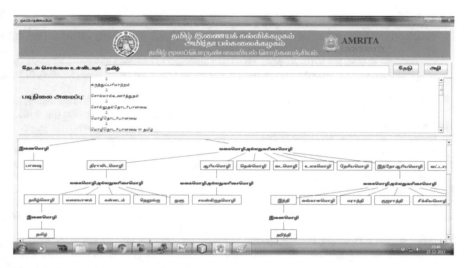

Fig. 16 Output of the user interface of OST

5 Conclusion

In this modern computerized era, searching and extracting knowledge and patterns are very important. A reliable and consistent intelligent system is always in demand in research, education and all other industries and institutions. We have developed a tool for information retrieval and a visualization system based on ontology methodology which gives the related entities with information for the user during searching process. A huge amount of data set has been collected and stored in an unstructured format. These data sets are converted to entities during generation of ontology tree and the generated tree is presented to the user.

References

1. Ananthapadmanaban, K. R., & Srivatsa, S. K. (2011, June). Personalization of user profile: creating user profile ontology for Tamilnadu Tourism. *International Journal of Computer Applications (IJCA), 23*, 42–47.
2. Athira, P. M., Sreeja, M., & Reghuraj, P. C. (2013, October). Architecture of an ontology-based domain specific natural language question answering system. *International Journal of Web & Semantic technology (IJWesT), 4*, 31–39.
3. Bhatt, B., & Bhattacharyya, P. (2012). Domain specific ontology extractor for Indian languages. In *Proceedings of 10th workshop on Asian Language Resources* (pp. 75–84). COLING: Mumbai.
4. Bolinger, D. (1967). Adjectives in English: Attribution and predication. *Lingua, 18*, 1–34.
5. Breslin, J. G., Passant, A., & Decker, S. (2009). *Social semantic web* (p. 58). Berlin: Springer.

6. Busa, F., Calzolari, N., Lenci, A., & Pustejovsky, J. (2001). Building a semantic lexicon: Structuring and generating concepts. In H. Bunt et al. (Eds.), *Computing meaning* (Vol. 2, pp. 29–51). Dordrecht: Kluwer Academic publishers.
7. Chaware, S., & Rao, S. (2010, August). Integrated approach to ontology development methodology with case study. *International Journal of Database Management Systems (IJDMS), 2*(3), 13–19.
8. Chaware, S., & Rao, S. (2012). Ontology supported inference system for Hindi and Punjabi. In *Proceedings of IEEE international conference on Technology Enhanced Education* (pp. 1–6).
9. Cruse, D. A. (1986). *Lexical semantics*. Cambridge: Cambridge University Press.
10. Cruse, D. A. (2000). *Meaning in language: An introduction to semantics and pragmatics*. Oxford: Oxford University Press.
11. Durga, A. K., & Govardhan, A. (2011, September). Ontology based text categorization- Telugu documents. *International Journal of Scientific & Engineering Research, 2*, 1–4.
12. Fellbaum, C. (1998). A semantic network of english verbs. In C. Fellbaum (Ed.), *WordNet: An electronic lexical database*. Cambridge: MIT Press.
13. Grover, P., & Chawla, S. (2014, May). Evaluation of ontology creation tools, *International Journal of Soft Computing and Engineering (IJSCE), 4*(2), ISSN: 2231–2307.
14. Jidge, P. (2017, October). Domain specific ontology extractor for Marathi language. *International Journal of Innovations & Advancement in Computer Science (IJIACS), 6*(10), ISSN 2347–8616.
15. Jidge, P., & Govilkar, S. (2016). A survey on ontology building methodologies and tools for Indian languages. *International Journal of Computer Science and Information Technologies (IJCSIT), 7*(2), 476–479.
16. Kaur, R., & Sharma, S. (2014, November). Pre-processing of domain ontology graph generation system in Punjabi. *International Journal of Engineering Trends and Technology (IJETT), 17*(3), 141–146.
17. Kaur, R., & Sharma, S. (2015, March). Techniques of ontology and its usage in Indian languages – a review. *International Journal of Computer Applications (0975–8887), 114*(5), 24–35.
18. Leech, G. N. (1974). *Semantics*. Harmondsworth: Penguin.
19. Lyons, J. (1977). *Semantics, volume I and II*. Cambridge: Cambridge University Press.
20. Maheswari, J. U., & Karpagam, G. R. (2010). A conceptual framework for ontology based information retrieval. *International Journal of Engineering Science and Technology, 2*, 5679–5688.
21. Mathur, I., Darbari, H., & Joshi, N. (2013). Domain ontology development for communicable diseases. *CS & IT-CSCP, 3*, 350–360.
22. Miller, G. A. (1991). *Science of words*. New York: Scientific American Library.
23. Miller, K. J. (1998a). Modifiers in WordNet. In C. Fellbaum (Ed.), *WordNet: An electronic lexical database* (pp. 47–67). Cambridge, MA: The MIT press.
24. Miller, G. A. (1998b). Nouns in wordNet. In C. Fellbaum (Ed.), *WordNet: An electronic lexical database* (pp. 23–46). Cambridge, MA: The MIT press.
25. Miller, G. A., Beckwith, R., Fellbaum, C., Gross, D., & Miller, K. J. (1990). Introduction to WordNet: An on-line lexical database. *International Journal of Lexicography, 3*, 235–244.
26. Nida, E. A. (1975). *Componential analysis of meaning: An introduction to semantic structure*. The Hague: Mouton.
27. Nida, E. A. (1975a). *Compositional analysis of meaning: An introduction to semantic structure*. The Hague: Mouton.
28. Nida, E. A. (1975b). *Exploring semantic structure*. The Hague: Mouton.
29. Pribbenow, S. (2002). Meronymic relationships: From classical mereology to complex part-whole relations. In R. Green, C. A. Bean, & S. H. Myaeng (Eds.), *The semantics of relationships: An interdisciplinary perspective* (pp. 35–50). Dordrecht/Boston/London: Kluwer Academic publishers.
30. Pustejovsky, J. (1995). *The generative lexicon*. Cambridge: MIT Press.

31. Rajendran, S. (1978). *Syntax and semantics of Tamil verbs* (Ph.D. Thesis). University of Pune, Pune.
32. Rajendran, S. (1983). *Semantic structure of Tamil vocabulary*. Report of the UGC sponsored postdoctoral work (in manuscript). Deccan College Post-Doctoral Research Institute, Pune.
33. Rajendran, S. (1995). Towards a compilation of a thesaurus for modern Tamil. *South Asian Language Review*, 5(1), 62–99.
34. Rajendran, S. (2001). *taRkaalat tamizc coRkaLanjciyam* [Thesaurus for Modern Tamil]. Thanjavur: Tamil University.
35. Rajendran, S. (2002, March). Preliminaries to the preparation of Wordnet for Tamil. *Language in India, 2*, 1. www.languageinindia.com
36. Rajendran, S. (2016). Tamil thesaurus to WordNet. In *Conference papers of 15th Tamil Internet Conference 2016. International Forum for Information Technology in Tamil*, September 2016, 1–9.
37. Rajendran, S., & Anandkumar, M. (2017, May). Visual Onto-thesaurus for Tamil. *Language in India, 17*, 5. www.languageinindia.com
38. Rajendran, S., & Baskaran, S. (2002). Electronic thesaurus for Tamil. In *Proceedings of the international conference on Natural Language Processing*. NCST: Mumbai.
39. Rajendran, S., & Baskaran, S.. (2006). *Tamizh mincoRkaLanjciyam* [Electronic thesaurus for Tamil]. Tamil Universtiy: Thanjavur.
40. Ramakrishnan, S., & Vijayan, A. (2014, April). A study on development of cognitive support features in recent ontology visualization tools. *Artificial Intelligence Review, Springer, Netherlands, 41*, 595–623.
41. Saraswathi, S., Siddhiqaa, A. M., Kalaimagal, K., & Kalaiyarasi, M. (2010, April). Bilingual information retrieval system for English and Tamil. *Journal of Computing, 2*, 85–89.
42. Slaughter, M. M. (1982). *Universal languages and scientific taxonomy in the seventeenth century*. Cambridge: Cambridge University Press.
43. Sowa, J. F. (1984). *Conceptual structures: Information processing in mind and machine*. Reading: Addison-Wesley Publishing Company.
44. Talita, P., Yeo, A. W., & Kulathuramaiyer, N. (2010). Challenges in building domain ontology for minority languages. In *Proceedings of IEEE international conference on Computer Applications and Industrial Electronics* (pp. 574–578).
45. Thaker, R., & Goel, A. (2015). A comparison of ontology based and keyword based Ouerv Processing System for Urdu Language. *International Journal of Research and Scientific Innovation (IJRSI), 2*(8), 19–23.

Machine Translation System for Translation of Malayalam Morphological Causative Constructions into English Periphrastic Causative

T. K. Bijimol, John T. Abraham, and D. Jyothi Ratnam

1 Introduction

Research in the area of Machine Translation has been increasing for several decades. Machine Translation is the sub-area of Natural Language Processing and it helps to convert one natural language into another [11]. Now, various Machine Translation systems are available on the internet that achieves certain level of accuracy, even though many translation issues need to be handled for better machine-translated output.

India demands translation of English into Indian languages and Indian languages into English because of India's numerous local languages. Google Translate is free software which provides translation services to translate English into Indian languages, Indian languages into English and Indian languages into other Indian languages. Each language has its own linguistic features and Google Translate is not able to identify linguistic features of any natural language. Difficulty in recognizing polysemic and synonymic nature of words is one main problem of available machine translation systems. In the first stage, this work conducts contrastive analysis of the structure of Malayalam morphological causatives and English analytical causatives. Research identified that Malayalam causative sentences have both 1st and 2nd causative verb forms. In order to develop a Malayalam–English Machine

T. K. Bijimol (✉)
Santhigiri College of Computer Science, Idukki, Kerala, India

J. T. Abraham
Department of Computer Science, Bharath Matha College, Kochi, Kerala, India

D. Jyothi Ratnam
Center for Computational Engineering & Networking (CEN), Amrita School of Engineering, Amrita Vishwa Vidyapeetham, Coimbatore, India

© Springer Nature Switzerland AG 2021
R. Kumar, S. Paiva (eds.), *Applications in Ubiquitous Computing*, EAI/Springer Innovations in Communication and Computing,
https://doi.org/10.1007/978-3-030-35280-6_11

translation system, this work translates Malayalam morphological causative positive (affirmative) sentence, negative and interrogative causative sentence into English by analysing structures of both Malayalam and English causative sentences.

According to surveys on machine translation methods and study on causative construction, causative marker attached to the verb decides the suffixes to be attached in the objects. But phrasal chunks being the main element for constructing the sentence, the case marker selection will not be feasible in the SMT method [2]. The translation of Malayalam causative sentences into English is not possible through modern techniques like Neural Machine Translation System (Google Translate) or Machine Learning approach because of unavailability of sufficient corpus or low resources. Unfortunately, sufficient quantities of Malayalam corpus containing causative sentences with its different morphosyntactical forms are not yet available. In this scenario, this work treats causative sentences with a well-developed rule-based direct approach. Direct approach is very economical that is suitable for the development of the translation system because only simple causative sentences are handled by this system. This method includes 6 modules which are: Tokenization, Suffix separation, POS (Parts of Speech) tagging, Causative verb identification, Malayalam-to-English word translation, word reordering and target sentence generation.

2   Morphosyntactic Properties of Malayalam Nouns

Linguistically, Malayalam is highly agglutinative in nature. Malayalam shows the morphological features like compounding, concatenation, derivation and inflections. The vocabulary of Malayalam is classified into two classes: *vachakam* (വാചകം) and *dyothakam* (ദ്യോതകം) [3, 4]. The *vachakam* (വാചകം) is again classified as the parts of speech (POS) categories like noun, verb and adjectives. The POS category like postpositions and adverbs comes under the title of *dyothakam* (ദ്യോതകം) [15, 16]. This category does not possess any individual meaning; it shows the relation between the words from different POS categories in a particular sentence.

The syntax and semantics of Malayalam is very unique which is that it does not demonstrate verb agreement for person, number and gender [15, 16]. Malayalam follows an unmarked word order of Subject-Object-Verb (SOV) [11, 12, 14]. The knowledge of the morphosyntactical features of Malayalam nouns and verbs is very essential for the NLP of Malayalam language, particularly in the context of machine translation. Malayalam shows morphological causativization. Morphological affixations are the significant identity of the morphological causation. Causative markers are denoted by the syntactic significance of Malayalam verb phrase (VP) [6]. Verbs imply the presence of three nominals, which are an Initiator, the Actor or performer, and an Object, and may be labelled causative verbs. Causative verbs always imply an Actor as well as an Initiator of the action performed by the Actor [6]. The usage

Table 1 Inflected forms of Malayalam nouns with neutral gender plural suffix

Noun	Animate		Inanimate
	Human	Non-human	
Singular + കൾ (kaL)	കുട്ടി (kuTTi) + കൾ (kaL) (child + plural suffix)	പട്ടി (paTTi) + കൾ (kaL) (dog + plural suffix)	പഴം + കൾ (kaL) (pazham) (fruits + plural suffix)
Plural	കുട്ടികൾ (kuTTikaL) (children)	പട്ടികൾ (paTTikaL) (dogs)	പഴങ്ങൾ (pazhangaL) (fruits)

of causative construction is very popular in Malayalam. The causative verbs in Malayalam follow the same case and tense, aspect and modality (TAM) agreement of transitive verbs and verbs of causative alternation.

An agglutinative language means a language in which the grammatical relations are expressed by affixes or suffixes added to the root or compounded with it [11]. Through compounding or affixation/suffixation, Malayalam nouns are capable of showing the morphosyntactic variations, and these various inflected forms possess 69 distinct semantics [11, 14–16]. The analyses of morphosyntactic properties of Malayalam nouns are very important which are listed below:

1. Malayalam nouns are inflected for number. The singular form of the noun is unmarked. The plural form is marked with three different plural suffixes. They are as follows:

   (a) കൾ (–kaL) [15, 16]
   (b) മാർ (–mAR) [15, 16]
   (c) കാർ (–kAR)/ആർ (aR) [15, 16]

(a) The suffix കൾ (–kaL) is used to produce the plural form of animate-human and non-human nouns and inanimate nouns [16] (Table 1).

(b) The plural suffix മാർ (–mAR) combine only with the human nouns.
    Examples:

    1. Singular noun Feminine + മാർ (–mAr): അമ്മമാർ (ammamAR) (Mothers) [11, 14–16].
    2. Singular noun Masculine + മാർ (–mAr) പുരുഷന്മാർ (purushanmAR) [11, 14–16].

(c) The plural suffix കാർ (–kAR) combining only with certain human nous, which possess certain features like the plural form does not reveal any gender specification of the members of the group [11, 14–16].
    Example: പോലീസുകാർ (polIsukAR) [policemen & women]

2. Malayalam nouns inflected for case.

Malayalam is a rich language with various suffixes. From these eight വിഭക്തിപ്രത്യയങ്ങൾ (*vibhakti prathyay*) case markers, three gender suffixes, and two number suffixes combine [11, 12, 14, 15] only with nouns and pronouns. When a case marker immediately precedes a noun, the ending consonant undergoes morphological variations [11, 14–16].

Examples:

Singular: കുട്ടി + എ (*kuTTi + e*) = കുട്ടിയെ (*kuTTiye*)
Plural: കുട്ടികൾ + എ = കുട്ടികളെ (*kuTTikLe*)
Singular: പുരുഷന് + ആല് (*purushan + Aall*) = പുരുഷനാല് (*purushanAl*)
Plural: പുരുഷന്മാര് + ആല് (*purushanmAr + Aall*) =
     പുരുഷന്മാരാല് (*purushanmArAall*)

3 Syntactic and Semantic Features of Malayalam Case System

The case system plays a very important role in the determination of the syntactic and semantic features of a sentence. The relation between nouns and verbs is known as കാരകം (*karakam*). Traditionally, Malayalam grammar follows the case system of Sanskrit. The case system in Malayalam is known as വിഭക്തിപ്രത്യയങ്ങൾ (*vibhakthiprkaraNam*) and case markers are known as the വിഭക്തിപ്രത്യയങ്ങൾ (*vibhakthiprathyyam*) [11, 14–16]. Case systems of Malayalam and Sanskrit are very similar. The Malayalam grammarians A.R. Rajaraja Varma and Seshagiri Prabhu follow the same naming system of Sanskrit for Malayalam case system. Malayalam and Sanskrit have seven cases and different cases are marked with different case markers. The sameness between Malayalam, Hindi and Sanskrit case system is very clear at various levels. Therefore, for the computing of Malayalam case system, we can easily adapt the perspectives of Paninian grammar [11, 14–16]. Malayalam case markers are used to show the relationship between nouns and other lexical items in a sentence; they always follow a noun only so they can be called as postpositions and they show many similarities between Hindi postpositions also. The specialties of Malayalam case system are:

1. A different വിഭക്തി (*vibhakti*) can be used for the same semantic relation with a given verb in a different sentence [11, 14, 15, 17].
2. Same വിഭക്തി (*vibhakti*) can be used with the same verb for two different semantic relations [11, 14, 15, 17]
3. In a noun group (NP), Malayalam case markers are placed immediately after the noun [11, 14, 15, 17]
4. Malayalam വിഭക്തി (*vibhakti*) is capable of bringing sense variations without affecting the syntax [11, 14, 15, 17].

5. The Malayalam വിഭക്തി (*vibhakti*) is capable of producing various semantic relations between the noun and verb in a particular syntax. (These factors are discussed in the section Syntax and Semantics of Malayalam വിഭക്തിപ്രകരണം (*vibhakti prakaranam*) [11, 14–16]
6. നിർദ്ദേശിക (*nirdeshika*) (nominative), പ്രതിഗ്രാഹിക (*prathigrahika*) (accusative), ഉദ്ദേശിക (*uddeshika*) (dative) and സംയോജിക (*samyojika*) (sociative) cases link the nouns to the basic structure of the sentence [11, 14, 16].

In the study of causative construction of Malayalam, നിർദ്ദേശിക (*nirdeshika*) (the nominal), പ്രതിഗ്രാഹിക (*prathigrahika*) (the accusative), ഉദ്ദേശിക (*uddeshika*) (the dative) പ്രയോജിക (*prayojika*) and (the instrumental) cases need more attention. We treated the case system of Malayalam with Paninian point of view.

Malayalam നിർദ്ദേശിക (*nirdeshika*) (the nominative case) is not having any particular case marker. This case marker notified the importance of the nominal entity, so if a nominal entity in nirdeshika, it never undergoes any kinds of morphological variations [2]. പ്രതിഗ്രാഹിക (*Prathigrahika*) (the accusative case) is marked with the suffix എ (*-e*), ഉദ്ദേശിക (uddeshika) (the dative case) is marked with ക്ക്/ന് (*'kku'/'nu'*) [11, 14, 16] and പ്രയോജിക (*prayojika*) (the instrumental case) is marked with two different markers like ആൽ (*–Aall*) and കൊണ്ട് (*-koNtu*) [11, 14, 16].

4 Morpho-Syntactic Properties of Malayalam Verbs

Malayalam is rich in verb. It borrowed a large number of verbs from Sanskrit and other languages like Tamil, Portuguese, etc. Traditionally, Malayalam verbs are classifieds into two groups (1) arjitham (borrowed) (2) swantham (own) [11, 14, 16]. In 1965, S. Kunjaan Pillai provided a comprehensive list of 2881 verbs divided into 16 classes. For denoting tense aspect modality [TAM], Malayalam verbs show inflections. Inflected forms of verbs are very common in Malayalam. This morphological uniqueness increases the complexity of the computing of Malayalam verbs in their NLP applications. The morphological structure of inflected verb forms varies from verb to verb based on their categories. The morphological categorization of verb depends on the ending consonants. Traditional grammarians morphologically categorize Malayalam verbs according to the ending sound of the tense form of the verb. They are:

1. Verb class of ഇ (*-i*) [15, 16]
2. Verb class of ച്ചു (*-icchu*) [15, 16]
3. Verb class of ത്തു (*-ththu*) [15, 16]

This categorization is equally applicable for intransitive, transitive and verbs of transitive alteration and causative verbs.

Three different suffixes ഇ (-*i*), ച്ചു (-*icchu*) and ത്തു (-*ththu*) are the past tense suffixes [15, 16, 19, 20]. The selection of the suffix depends of the category of the verb. The present tense suffix is ന്നു (-*unnu*) and this tense possesses many language-specific features. The future tense of Malayalam expresses the tense, aspect and molality (TAM) suffixes.

5   The Syntax and Semantics of Malayalam Causatives

We can define that the causation or causativization adds an additional participant to an event: a new causer. A causative construction with a causative verb derived from a basic verb through a regular morphological process, for example, affixation is called morphological causative construction [12, 18].

In Malayalam, some verbs possess 1st and 2nd causative forms and some others have only one form. പ്പി (-*ppi*) and are the ക്കി/ത്തി ('*kki/ -ththi*') different causative suffixes of Malayalam. The causative suffix ക്കി/ത്തി ('-*kki/ -ththi*') *is denoted as 1st causative and* പ്പി (-*ppi*) is denoted as the 2nd causative of Malayalam causative verbs [18]. A transitive verb possesses ക്കി/ത്തി (*kki/ -ththi*) and പ്പി (-*ppi*) *forms* and, an intransitive verb possesses the only one causative alternated form with പ്പി (-*ppi*) [15, 16].

6   Translation of the Sentence with Verbs of Causative
Alternation and Causatives

(a) *Type one*

   1a. പാത്രം ഉടഞ്ഞു
   (*pAthramudanju*)
   ET: The vase broke [9].
   1b. അവൻപാത്രം ഉടച്ചു
   (*avanpAthramudacchu*)
   ET: He broke the vase [9].
   1c. അവന് പാത്രം ഉടപ്പിപ്പിച്ചു
   (*avanpathramudappicchu*)
   ET: He had/made the vase broken
   1d. അവൻരാമനെ കൊണ്ട് പാത്രംഉടപ്പിച്ചു
   (*avanramanekondupAthramudappicchu*)
   ET: He had/made Rama the vase broken.

The above given examples reveal the following facts:

1. The main verb of sentence 1a is intransitive.
2. The main verb of sentence 1b is the causative alternation of the intransitive verb ഉടയുക (*udayuka*): (broke)

3. The main verb in 1c is the causative form of the verb ഉടയൽ (*udayul*) (broke); the recipient of the causative event is an inanimate object, so it is marked with the accusative case with null element [9].
4. The 1d, there is no mediator cause, so the first cause is marked with a compound case marker യെകൊണ്ട് (*-yekondu*) because this animate object acts as an instrument. The recipient is an inanimate object so it is marked with accusative case marker with null element.

(b) *Type two*

> 2a. അമ്മ കുട്ടിയെ കുളിപ്പിപ്പിച്ചു
> (ammAkuTTiyekuLippicchu)
> ET: Mother bathed the child.
> 2b. അമ്മകുട്ടിയെസീതയെകൊണ്ടുകുളിപ്പിച്ചു
> (ammAkuTTiyesIthyekoNdukuLippicchu)
> ET: Mother caused Sita to bathe the child

The above two sentence 2a and 2b reveal the following facts.

1. The main verb in sentence 2a is transitive.
2. The main verb in sentence 2b is the causative form of the main verb. The recipient of the causative event is an animate object, so it is marked with accusative case marker with എ (*-e*) and the agent is marked with instrumental case marker with യെകൊണ്ട് (*-yekondu*).

(c) *Type three*

3. അവന് രാമനെകൊണ്ട് കുട്ടിക്ക് പൈസ കൊടുപ്പിച്ചു
(avan rAmanekondukuTTiykku paisa koduppicchu)
ET: He caused Ram to give money to the child

In this causative sentence, the recipient is the child, is an animate object. Here the verb belongs to the class 'give' so the recipient is marked with dative case with the case marker ക്ക് (*kku-*)

(d) *Type four*

> 4a. അവൾകേക്ക് ഉണ്ടാക്കിച്ചു
> (*avaLkekkuundaakkicchu*)
> ET: She had the cake made.
> 4b. അവൾബേക്കറിക്കാരനെകൊണ്ട് കേക്ക് ഉണ്ടാക്കിപ്പിച്ചു
> (avaLbekkaRikkAranekondukekkuundakkippicchu)
> ET: She had the baker made the cake.

The causative verb form of the sentence 4a is 1st casual form and 4b is 2nd casual form.

For the analysis of Malayalam causative sentences, we follow the Paninian perspective of grammar. The case systems as well as the determination of subject,

1st cause and 2nd cause, recipient, causative event/events causational situations or 1st causative and 2nd causative are in accordance with Panini. In Paninian grammar, he uses certain technical terms for the identification of the subject cause/causes. The real subject is termed as പ്രയോജകകർത്ത (*Prayojakakartha'*) [4, 5]; it is marked with nominative cause. The mediator cause is termed as മത്ൃത്ഥകർത്ത (*MathyasthakarthAvu*) *and* it is marked with instrumental case and the first cause is termed as പ്രയോജകകർത്ത (*PryojyakarthAvu*) and is marked with accusative case/dative case.

7   English Periphrastic or Analytical Causative Constructions

In periphrastic causative construction, causation is indicated by the help of an auxiliary verb, which occurs as part of main verb [7]. The auxiliary verbs 'have', 'make', 'get', 'let', and 'help' are used for some grammatical functions in English language. In addition, it is capable of producing some linguistic functions, which are the 'causing verbs' in periphrastic causative constructions. For example:

1. I had my car washed

Two-verb construction is the main features of a periphrastic causative sentence in which the first verb represents the predicate of causation and main verb or second verb indicates a predicate of effect. The whole verb phrase together expresses the semantics of the sentence.

2. Sita made the baby dance

In this sentence, 'Sita' is predicate of causation and dance is the predicate of effect.

3a. He made her cry.
3b. He made her write the report.

In examples 3a and 3b, 'he' is the subject which is the causer – 'her' is the lexical item which performs the action and so are termed as 'causee' of the causative event. 'Cry' and 'write' are the causing event and the noun phrase (NP/PRP/PP) 'her', 'the report' are termed as the 'affectee'. In this construction, the auxiliary verb 'made' is the causative verb. The major constituents of a causative construction are the 'causer', the 'causee', the 'affectee', 'causative verb' and the 'caused event' [4, 7, 12, 18].

Periphrastic English causative constructions are divided into two types which are impersonal causatives and interpersonal causatives [7, 12, 18]. In the example, 'Sita made Ravi feed the apple to Krishna', Ravi is the direct cause and Sita is indirect cause, both of which exist in one sentence and it can be considered as interpersonal causative construction. But in example, 'Sita fed the apple to Krishna', there is no indirect cause and such type of sentence is called impersonal causative constructions.

8 Working Model of Malayalam-to-English Causative MT System

The proposed system is implemented using Python Programming language. Figure 1 shows the working model of the rule-based Malayalam–English machine translation system for causative construction.

8.1 Tokenization and Suffix Separation

Tokenization is the process in which sentences are partitioned into small individual units named tokens [24].

Space is a delimiter for tokenization in many languages like English, but it is not possible with languages which have no clear boundaries. Malayalam is one among them and it is necessary to get morphological and lexical information in order to tokenize Malayalam sentences. In the tokenization process, the input

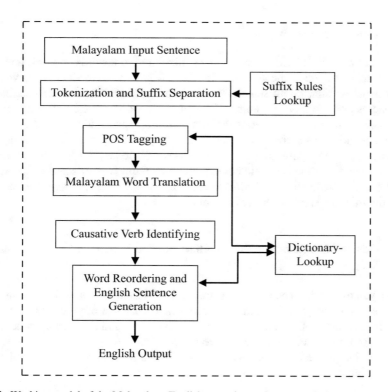

Fig. 1 Working model of the Malayalam–English causative sentence translation system

sentence is divided into individual tokens such as words or punctuation marks or numbers. Following examples show the output of tokenization function for the given Malayalam Input (MI) sentence.

Example MI 1.അവൾബേക്കറിക്കാരനെകൊണ്ട്കേക്ക് ഉണ്ടാക്കിപ്പിച്ചു.

$$(avaL\ bekkaRikkAranekkoNdu\ kekku\ undakkippichu)$$

Output: 'അവൾ', 'ബേക്കറിക്കാരനെകൊണ്ട്', 'കേക്ക്', 'ഉണ്ടാക്കിപ്പിച്ചു'

$$\left({}^{'}avaL^{'},\ {}^{'}bekkaRikkAranekkoNdu^{'},\ {}^{'}kekku^{'},\ {}^{'}undakkippichu^{'} \right)$$

Malayalam is an unsegmented language and ബേക്കറിക്കാരനെ കൊണ്ട് (bekkaRikkAranekkoNdu) is inflected word or unsegmented word which can be divided into individual morphemes. In Malayalam, an inflected word can be divided with the help of the suffix separation rules.

8.1.1 Suffix Separation

Malayalam word is formed by combining suffixes and root words or word stem [23]. Word stem + suffixes is the syntax of inflected words of Malayalam. The two main components of Malayalam sentences are noun and verb [22]. So verb stems and noun stems are two types of stems. Verbs are infected by tense, aspect and mood, and thus, verbal forms are added with different tense, aspect and mood suffixes. Case markers are added as suffixes for nouns. Suffix separation method uses a root word dictionary for identifying the valid stem, a suffix dictionary for finding out a valid suffix and sandhi rules. The output of example 1 contains two inflected words, which are 'ബേക്കറിക്കാരനെ' (bekkaRikkArane) and ഉണ്ടാക്കിപ്പിച്ചു (undakkippicchu). Here, the first one is a noun and the second, a verb.

8.1.2 Design of Malayalam Tokenization and Suffix Separation Algorithm [13]

In Malayalam, different Sandhi rules are used for joining two words to generate a new word. The original form of the word is altered when applying these Sandhi rules. The 'sounds' of the end syllable of the first word and the start syllable of the second word are observed and rules are applied based on it. The Sandhi rules are given based upon the word ends with a vowel (swaram) or a consonant (vyanjanam) in Malayalam sentence.

Table 2 Noun suffix table for causative construction

Case	Suffix	Abbreviated form
Nominative	Null	Null
Accusative	എ (-e)	െ
Instrumental	കൊണ്ട് (*koNdu*)	ൊണ്ട്

(a) *Noun Suffixes*

Case markers are deeply analysed and observed that nominative, accusative and instrumental case markers are only relevant for noun suffixes of the causative sentence construction that can be shown in Table 2. In a causative construction, the instrumental case marker is in a form of a compound marker 'യെകൊണ്ട്' ('-*yekoNdu*') with the combination of accusative marker 'എ' ('-*e*') and instrumental case marker 'കൊണ്ട്' ('*koNdu*'). Suffixes in the inflected words are not in its original form, so types of suffixes are to be found out at first. The accusative case is considered the inflected word 'കുട്ടിയെ' ('-*kuttiye*') can be split as 'കുട്ടി' ('-*kutti*') + 'എ' ('-*e*'). Here, 'എ' ('-*e*') is present in an abbreviated form as 'െ'. These abbreviated forms are used to identify the suffixes. Examples of three case markers with its abbreviated forms are given in Table 2.

(b) *Verb Suffixes*

Inflected verb structure varies from verb to verb. It depends upon their categories and the categorization is based on the ending consonants. For example, past tense forms of the main verbs categorized based on the ending sounds are: 'ഇച്ചു' (*icchu*) and 'പ്പിച്ചു' (*ippicchu*) are affirmative sentence impersonal and affirmative sentence interpersonal form of suffixes. 'ഇച്ചില്ല' (*icchilla*) and 'പ്പിച്ചില്ല' (*ippicchilla*) are impersonal and interpersonal negative sentence form of suffixes. 'ഇച്ചോ?' (*icchO?*) and 'പ്പിച്ചോ?' (*ippicchO?*) are impersonal and interpersonal interrogative sentence form of suffixes. All available categories of past, forms of affirmative, negative and interrogative sentence suffixes and their abbreviated forms are given in Table 3.

In order to separate the suffixes of Malayalam words, algorithm applies the Sandhi rules in the reverse direction. It means, word structure is analysed from right to left, identified the split_letter and split the words according to the suffix separation rules [21]. A lexical database which contains noun and verb roots in Malayalam is used for suffix separation. Some words which belong to accusative case have equivalent translations in English without their root form. For example, the word 'അവനെ' ('*avane*') has no root forms in Malayalam and it has direct translation word in English. The word 'him' is the English word which is semantically equivalent to the Malayalam word 'അവനെ' ('*avane*'). These types of words are categorized into split_exception category 'അവളെ' ('*avale*') which is also the part of accusative case that is stored as split_exception category.

Table 3 Verb suffix table for positive, negative and interrogative forms of past tense forms of causative construction in Malayalam

Tense	Types of sentence (Positive/negative/interrogative)	Causative impersonal affirmative forms of main verb suffixes in Malayalam 1st causative form	Abbreviated forms of suffixes 1st causative form	Causative personal affirmative forms of main verb suffixes in Malayalam 2nd causative form	Abbreviated forms of suffixes in 2nd causative form	English causative form
Past tense	Positive	ഇച്ചു (*icchu*)	ച്ചു (*icchu*)	ഇപ്പിച്ചു (*ippicchu*)	പ്പിച്ചു (*ippicchu*)	Had, got, made
	Negative	ഇച്ചില്ല (*icchilla*)	ച്ചില്ല (*icchilla*)	ഇപ്പിച്ചില്ല (*ippicchilla*)	പ്പിച്ചില്ല (*ippicchilla*)	
	Interrogative	ഇച്ചില്ലേ? (*icchillai?*)	ച്ചില്ലേ? (*icchillai?*)	ഇപ്പിച്ചില്ലേ? (*ippicchillai?*)	പ്പിച്ചില്ലേ? (*ippicchillai?*)	

8.1.3 Design of Suffix Separation Rules

Split_letter is the basic element for design of rules in suffix separation. An input word is scanned from right-hand side and identified a matching suffix from the suffix table. Previous letter is taken as split_letter and applies the rules using suffix table, which is given in the table. As an example, in the word 'കോമളനെ', Prev_Word_ന denotes the sub string 'കോമള.' Table 4 illustrates the suffixes, suffix rules and example words.

Output

കുട്ടിയെക്കൊണ്ട് = കുട്ടിയെ + ക്കൊണ്ട്
കുട്ടിയെ = കുട്ടി + എ
കഴുകിപ്പിച്ചു = ഇപ്പിച്ചു + കഴുക്

8.2 Part of Speech (POS) Tagging

POS tagging is the process of assigning tags to individual units of a sentence [8]. It explains how a word is used in a sentence. Tag set used in this work can be seen in Table 5.

Noun and verb dictionary contains one tag entry which contains POS tag information of all tokens or morphemes. Read the dictionary entry that assigns POS tag to given morphemes.

Sample Output:

PN(അവൾ) (*avaL*) N(കുട്ടി) (*kutti*) NA(എ) (*e*) NA(കൊണ്ട്) (*koNdu*) PN(എന്റെ) (*eNte*) N(കാർ) (*cAr*) V(കഴുക്) (*kazhuk*) VA(ഇപ്പിച്ചു) (*ippiccu*)

8.3 Malayalam Causative Verb Identifying and Processing

The working principle of the newly developed system for Malayalam-to-English translation is given below:

The morphological analysis of Malayalam input sentence is the initial processing of the Malayalam-to-English translation. The analysed input comes under the causative verb identifier; here, the system identifies whether the given input is causative or not; otherwise, the system inputs the next sentence for processing. If the sentence is causative, system identifies whether the given sentence is 1st causative or 2nd causative form and the tense. If yes, then given input is swept into the Malayalam-English Word Translation Module. The translated linguistic elements are arranged into the analytical causative sentence pattern of target language, English. If the input sentence belongs to 1st causative, the system will choose the impersonal analytical causative sentence pattern. If the input Malayalam

Table 4 Suffix table

Rule No.	Split letter	Suffix_Id	Suffix symbol	Suffix separation rules	Functions	Example	Root extracted
1	ന(n)	ai	ൈ	Prev_Word_നൈ + suffix	Can select words ending with ൻ	മകനെ	മകൻ
2	ല(l)	ai	ൈ	Prev_Word_ലൈ + suffix	Can select words ending with ൽ	ലാലെ	ലാൽ
3	ള(L)	ai	ൈ	Prev_Word_ളൈ + suffix	Can select words ending with ൾ	കുട്ടികളെ	കുട്ടികൾ
4	റ(R)	ai	ൈ	Prev_Word_നൈ + suffix	Can select words ending with ർ	അമറിനെ	അമർ
5	ണ(N)	ai	ൈ	Prev_Word_നൈ + suffix	Can select words ending with ൺ (Nu)	കിരണിനെ	കിരൺ
6	For consonants like ക, ഖ (k, kh) etc. use യ as split letter	ai	ൈ	Prev_Word_യൈ+ suffix	Can select words ending with എ and ഇ (ai&e)	കനകരയ	കനക
						സീതയെ	സീത
						കുട്ടിയെ	കുട്ടി
7	For words end with അം and ഉ (aM& u) sound, with 'semiu' (samvrutha ukAram)	am	ൊം	Prev_Word_Sൊം + suffix	Can select words ending with അം sound	ഒട്ടകത്തെ	ഒട്ടകം
		ai	്	Prev_Word_SL+് + suffix	Can select words ending with 'semi-u' (samvrutha ukAram)	സുഹൃത്തിനെ	സുഹൃത്ത്
8	ക്ക	kk	ൊാണ്ട്	Prev_Word_SL + suffix	Can select words ending with case marker ക്കൊണ്ട്	അവനെക്കൊണ്ട്	അവനെ

Table 5 Malayalam POS tags

Sl. no.	Tag description	Tag	Sl. no.	Tag description	Tag
1	Noun	N	7	Adverb	PAV
2	Verb	V	8	Adverbial suffix	ADVA
3	Postposition	NA	9	Noun clause suffix	NCA
4	Adjective	PA	10	Infinitive	INFA
5	Plural suffix	PL	11	Verbal suffix	VA
6	Adjectival suffix	ADJA	12	Pronoun	PN

sentence belongs to 2nd causative sentence pattern, the system chooses the personal analytical causative sentence pattern:

The following examples show the selection of pattern of English personal/impersonal periphrastic causative sentences.

Input sentence: അവൾകുട്ടിയെകൊണ്ട് റിപ്പോർട്ട് എഴുതിപ്പിച്ചു

(*avaL kuttiye koNdu riportt ezhuthippicchu*)

The sentence pattern of Malayalam causative sentence is:

Prayojaka kartha + prayojaya kartha (animate object) + prayojika vibhakti pratyay (instrumental case marker) + patient (inanimate object).

Based on the rules, the system chooses the interpersonal analytical causative [1] sentence pattern of English in past tense form.

ET: She had the child write the report.

Figure 2 shows the flow chart of causative sentence identifying and processing module.

8.3.1 Algorithm

Begin

1. Read the sentence
2. Scan the sentence and identify the Causer (Subject), Cause (Secondary object), Patient (primary object) and main verb form in the input sentence. Perform steps (a)–(c).

   (a) If sentence contains cause and patient and the case marker 'കൊണ്ട്' ('-*kondu*') exists with cause, then the sentence is in 2nd causative form.
   (b) Else sentence is in 1st causative form
   (c) Else print message 'Sentence is not causative' and go to step 1.

End

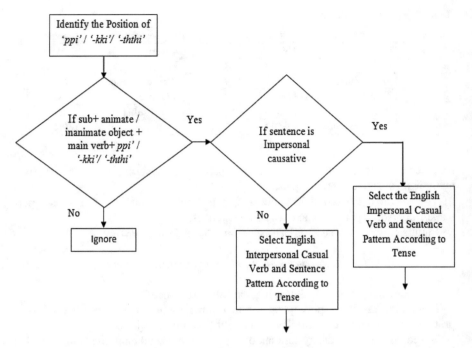

Fig. 2 Flow chart of rule-based Malayalam–English causative sentence processing

8.3.2   Tense Forms

There are nine different Malayalam tenses handled in this work that are listed in Table 7.

8.4   Malayalam Word Translation Function

It is required to keep a dictionary for storing root words and with various inflectional forms of causative verbs of Malayalam and the main verbs and its various forms of English verbs and different forms of causative 'have'. Using Malayalam–English bilingual noun dictionary, the system selects equivalent English words if Malayalam word is available in the dictionary. Also, use the verb dictionary for choosing equivalent causative verb forms. Malayalam–English bilingual noun and verb dictionaries are two dictionaries used for Malayalam–English MT systems. Noun dictionary contains Malayalam noun words and their equivalent English words. It also contains Parts of Speech information. Verb dictionary has causative verb root forms along with different past/present/future tense verb forms [10].

8.5 Reordering and English Sentence Generation Module

In this module, the system applies the linguistic rules of English for rearranging the translated Malayalam words. Malayalam follows the SOV order, but English has SVO order. The sentence order and pattern of the Malayalam is:

(a) *Malayalam (1st Causative Form) Sentence Pattern:*

Prayojaka kartha + patient (inanimate/animate object) + Main verb in first causative form.

(b) *English (Impersonal Form) Sentence Pattern:*

'Subject + causative verb (have/make/get) + primary object (inanimate object) + English main verb in past form.

Sample Output:
Malayalam Input (MI): ഞാൻഎന്റെമുറികഴുകിച്ചു.

(njAn eNte muRi kazhukichchu.)

ET: I had my room cleaned.

(c) *Malayalam 2nd Causative Form Sentence Pattern:*

Subject + causative verb (have/make/get) + agent (animate object) + English main verb in its root form + inanimate object.

(d) *English (Interpersonal Form) Sentence Pattern:*

'Subject + causative verb (have/make/get) + agent (animate object) + English main verb in its root form + inanimate object.

MI: ജോണി തന്റെ മകനെ കൊണ്ട് സൈക്കിൾ തുടപ്പിച്ചു.

(jOni thnte makane kondu saikkiL tudappichchu.)

ET: John had/got/made his son wipe the cycle.

9 Result and Discussions

The analysis phase uses the test corpora which are collected from literature, news and other digital repositories such as review books and articles of various domains. Data selection was performed randomly and care had been taken to make sure that data contain verity of sentence constructs. All possible simple sentences and some complex sentences are incorporated in the test set. Following are the translations generating from the proposed system.

Malayalam Input:

1. ഞാൻഎന്റെമുറികഴുകിച്ചു. (*njAn eNte muRi kazhukicchu.*)

   System Output (SO): I had my room cleaned.

   Google Translate Output (GTO): I washed my wound.

2. ജോണി വേലക്കാരനെ കൊണ്ട് സൈക്കിൾതുടപ്പിച്ചു. (*jOnivelakkarane kondu saikkiL tudappicchu.*)

   SO: John had servant wipe the cycle.

   GTO: Johnny stopped the bike with his son.

3. (ജോണി തന്റെ മകനെ കൊണ്ട് സൈക്കിൾതുടപ്പിച്ചില്ല) (*jOni thnte makane kondu saikkiL tudappicchilla.*)

   SO: John did not have his son wipe the cycle.

   GTO: Translation error

4. എന്ത് ഞാൻഎന്റെ മുറി കഴുകിച്ചില്ലേ? (*enthu njAn ente muRi kazhukicchillai?*)

   PTO: Did I have my room cleaned?

   GTO: What have I washed my room?

9.1 Performance Evaluation

Human evaluation, round trip translation evaluation and automatic evaluation are the available methods. Human evaluation is the most reliable method to compare quality of various machine translation systems output [24]. A language expert is required to correctly evaluate the translated sentences. The professional human translators are the best, but it will take months to finish and lead to high translation cost, effort, non-reusability and subjectivity. Adequacy and fluency are two human evaluation criteria for the evaluation of MT output [3].

Adequacy is an accuracy measure that refers to a gold standard translation to evaluate how much of the meaning is expressed in output text. Fluency measures the output readability, which means how much translation is grammatically well informed. A 5-point scale can be used to rate the adequacy and fluency [3, 8] as listed in the Table 6. Human evaluation is accurate and safe, but it is expensive and time consuming.

Table 6 Five-point scale to rate the adequacy and fluency

Adequacy		Fluency	
5	All	5	Flawless English
4	Most	4	Good English
3	Much	3	Non-native English
2	Little	2	Disfluent English
1	None	1	Incomprehensible

9.2 Performance Evaluation of Malayalam-to-English Machine Translation

Prototype translator uses 100 simple sample sentences of each tense category (Fig. 3; Table 7). Adequacy and fluency are the two measures to evaluate the translation in human evaluation method (Fig. 4; Table 8).

9.3 Discussions

The newly developed Malayalam-to-English MT system is compared with Google Translate. The Google translator translated all the main verbs in their transitive

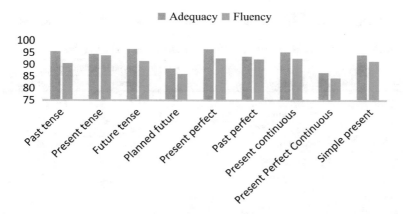

Fig. 3 Graph depicting performance of Malayalam-to-English prototype translator

Table 7 Accuracy testing of Malayalam-to-English prototype translator

Sl. no.	Tense forms	Prototype translator output	
		Adequacy	Fluency
1	Past tense	95.55	90.59
2	Present tense	94.45	93.90
3	Future tense	96.55	91.56
4	Planned future	88.34	86.04
5	Present perfect	96.55	92.77
6	Past perfect	93.45	92.345
7	Present continuous	95.32	92.67
8	Present perfect continuous	86.54	84.34
9	Simple present	94.12	91.46
Average		93.43	90.63

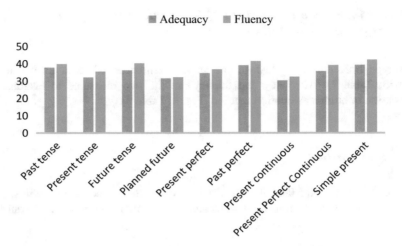

Fig. 4 Graph depicting performance of Google Translate Malayalam–English translation

Table 8 Accuracy testing of Google Translate Malayalam–English translations

Sl. no.	Tense forms	Prototype translator output	
		Adequacy	Fluency
1	Past tense	37.78	39.91
2	Present tense	32.12	35.56
3	Future tense	36.21	40.34
4	Planned future	31.56	32.23
5	Present perfect	34.6	36.78
6	Past perfect	39.11	41.56
7	Present continuous	30.34	32.56
8	Present perfect continuous	35.67	39.222
9	Simple present	39.23	42.33
Average		35.18	37.83

form only; it means that Google translator failed to identify the causative sense of Malayalam source sentences. Impersonal or 1st causative and interpersonal or 2nd causative sentences are used for testing. The system gave quality output for simple sentences with nearly an accuracy of 92.0341% in Malayalam-to-English translation system but Google Translate has only 36.508% accuracy with same input. Most of the sample sentences are in Subject-Object-Verb format. In the context of long sentences, proposed MT system failed to recognize the exact sentence patterns of source and target languages.

10   Conclusion and Future Enhancements

The output of the proposed system was compared with the output of Google Translate and identified that the Malayalam-to-English MT system performs translations with outstanding accuracy. This translator provided the quality output for simple sentences, but it is unable to handle other sentence patterns like complex and long sentences. In future, it can be extend to handle long causative sentences for better result. The bilingual dictionary has to be enhanced for handling all types of words in Malayalam. Quality results were obtained with simple sentences, but system needs to be improved for complex and lengthy sentence translation.

References

1. Anju, E. S., & Manoj Kumar, K. V. (2014). Malayalam to English machine translation: An EBMT system. *IOSR Journal of Engineering (IOSRJEN), 4*(1). www.iosrjen.org.
2. Auersperger, M. (2012). *English causative constructions with the verbs have, get, and make, and their Czech translation counterparts.*
3. Banchs, R. E., D'Haro, L. F., & Li, H. (2015). Adequacy – Fluency metrics: Evaluating MT in the continuous space model framework. *IEEE/ACM Transactions on Audio, Speech and Language Processing, 23*(3), 472–482.
4. Begum, R., & Sharma, D. M. (2010). *A preliminary work on Hindi causatives*. In Proceedings of the eighth workshop on Asian language resources, pp. 120–128.
5. Bharati, A., Chaitanya, V., Sangal, R., & Ramakrishnamacharyulu, K. V. (1995). *Natural language processing: A Paninian perspective* (pp. 65–106). New Delhi: Prentice-Hall of India.
6. Gopinathan Nair, B. (2008). *Collected papers on Malayalam language and linguistics.* Thiruvanathapuram: International School of Dravidian Linguistics.
7. Kemmer, S., & Verhagen, A. (1994). The grammar of causatives and the conceptual structure of events. *Cognitive Linguistics, 5*(2), 115–156. https://doi.org/10.1515/cogl.1994.5.2.115.
8. Koehn, P., & Monz, C. (2006). Manual and automatic evaluation of machine translation between European languages. In *Proceedings of the workshop on statistical machine translation*. New York: Association for Computational Linguistics.
9. Levina, B., & Hovav, M. R. (2016). *A preliminary analysis of causative verbs in English.* Department of Linguistics, Northwestern University, Sheridan Road, Evanston, IL, USA, Department of English, Bar Ilan University, Ramat Gan, Israel.
10. Macketanz, V., Avramidis, E., Burchardt, A., Helcl, J., & Srivastava, A. (2017). Machine translation: Phrase-based, rule-based and neural approaches with linguistic evaluation. *Cybernetics and Information Technologies, 17*(2), 1–18.
11. Mitkov, R. (2003). *The Oxford handbook of computational linguistics*. Oxford: Oxford University Press.
12. Nadathur, P. (2017). *Causative verbs: Introduction to lexical semantics*, pp. 1–15.
13. Nair, L. R., David Peter, S., & Ravindran, R. P. (2012). Design and development of a Malayalam to English translator: A transfer based approach. *International Journal of Computational Linguistics (IJCL), 3*(1), 1–11.
14. Narayanapillai, K. S. (1995). *Adhunika Malayalam vyakaraNam*. Thiruvananthapuram: Kerala Bhasha Institute.
15. Rajaraja Varma, A. R. (2006). *Keralapanineeyam* (8th ed.). Kottayam: DC Books.
16. Ram, K. V. (2001). *Ramkumarinte Sampourna Malayala Vyakaranam*. Thiruvananthapuram: Siso Books.

17. Ratnam, J. D., Kumar, M. A., Premjith, B., Soman, K. P., & Rajendran, S. (2018). Sense disambiguation of English simple prepositions in the context of English–Hindi machine translation system. In S. Margret Anouncia & U. Wiil (Eds.), *Knowledge computing and its applications*. Singapore: Springer.
18. Ratnam, J. D., Soman, K. P., Biji Mol, T. K., & Priya, M. G. (2019). Translation equivalence for English periphrastic causative constructions into Hindi in the context of English to Hindi machine translation system. In R. Kumar & U. Wiil (Eds.), *Recent advances in computational intelligence. Studies in computational intelligence* (Vol. 823). Cham: Springer.
19. Ravi Sankar, S. N. (2012). A grammar of Malayalam. Journal of Language in India 12, 2012
20. Rev, R. C., & Nair, S. K. (1993). *Draavida bhaashaa vyaakaranam (part one)* (3rd ed.). Thiruvananthapuram: The State Institute of Languages.
21. Sebastian, M. P., Sheena Kurian, K., & Santhosh Kumar, G. (2010). *A classification of Sandhi rules for suffix separation in Malayalam*. In 38th All India conference of dravidian linguists.
22. Shahana, I. L., & Sharafudheen, K. A. (2014). Relative study on Malayalam-English translation using transfer based approach. *International Journal of Computing and Technology, 1*(2), 24–29.
23. Somers, H. (2005). *Round-trip translation: What is it good for?* In Proceedings of the Australasian language technology workshop 2005, pp. 127–133.
24. Steven, B., Edward, L., & Ewan, K. (2009). *Natural language processing with python*. Sebastopol: O'Reilly Media Inc..

Hybrid Machine Translation System for the Translation of Simple English Prepositions and Periphrastic Causative Constructions from English to Hindi

D. Jyothi Ratnam, K. P. Soman, T. K. Bijimol, M. G. Priya, and B. Premjith

1 Introduction

Machine Translation (MT) is one of the applications of Computational Linguistics (CL) and Artificial Intelligence (AI). When a text in one human language is translated into another human language by a machine, it is called Machine Translation. Machine Translation is one of the applications of Natural Language Processing (NLP).

All natural languages are highly complex, and each has its own language specificity. Most of the lexical items in all natural languages exhibit polysemy and synonymy. The syntactic complexity of languages, the variations in syntactic constructions found in the languages, different semantics of same and variant syntactic constructions, and the polysemic and synonymic nature of lexical items of natural languages; still remain unresolved problems in the context of Machine Translation (MT). Translation is a series of steps which involves the replacement of human thoughts expressed in one natural language elements (Source Language, SL) into its equivalent natural language elements in an another natural language (Target Language, TL). The process of translation involves:

D. Jyothi Ratnam (✉) · K. P. Soman · B. Premjith
Center for Computational Engineering & Networking (CEN), Amrita School of Engineering, Amrita Vishwa Vidyapeetham, Coimbatore, India

T. K. Bijimol
Santhigiri College of Computer Science, Idukki, Kerala, India

M. G. Priya
Department of English & Humanities, Amrita Vishwa Vidyapeetham, Coimbatore, Tamil Nadu, India

© Springer Nature Switzerland AG 2021
R. Kumar, S. Paiva (eds.), *Applications in Ubiquitous Computing*, EAI/Springer Innovations in Communication and Computing, https://doi.org/10.1007/978-3-030-35280-6_12

Comprehension: the ability to understand and interpret spoken and written language

Formulation: putting together the linguistic elements in appropriate syntax to express the context of the source text

The nature of expressiveness is got due to the polysemic and synonymic nature of the lexical items. The lexical items are classified as functional words and content words. The major part of the lexicon contains the content words. The functional words are few in number or they are countable. The functional words play a vital role into determining the syntax and semantics of the sentences in a particular natural language. The functional words like pre/postpositions, auxiliary verbs and helping verbs are highly polysemic and synonymic in nature [7]. They perform many grammatical functions and build different types of sentences. They also show one-to-many and many-to-one translation equivalence in different languages [7]. Identification of the sense and the grammatical functions of the functional words in one natural language and finding translation equivalence for them in another natural language are the most demanding tasks in MT. Recognizing the exact grammatical functions of the functional words of the Source Language (SL) and finding their translation equivalence in Target Language (TL) by a machine leads to the betterment of the quality of output of the existing MT systems. India is a multi-lingual as well as one of the fastest-growing countries in the world. A well-established communication system and high-quality English-to-Indian Languages (EILMT), Indian Language to another Indian language (ILILMT), and one Indian Language to English machine translation systems (ILEMT) are the need of our time. A number of attempts have been made in India to build machine translation systems for transfer of data from English-to-Indian languages, one Indian language to another Indian language, and Indian languages to English. Hindi and English are two important official languages of the Government of India. India is a country with 22 official languages from five language families; hence, the use of Hindi is a challenging task even for Indian citizens speaking other languages. Of late, Google Translate has become popular among the internet users who wish to communicate and translate text from English to Hindi, Hindi to other Indian languages, Hindi to English, and other Indian languages to English also. While using Google Translate (GT)/Bing Translate (GT/BT)/Technology Developed for Indian languages (TDIL) for Hindi translation, if it is fed with simple sentences from English, the Google Translate (GT/BT/TDIL) mostly gives the correct Hindi translation. If the input sentences are long, the GT/BT/TDIL mostly fails to give the correct translation. It has been observed that when the GT/BT/TDIL is fed an English sentence with language specificity of English language, the machine is unsuccessful to capture the unique features of the source sentence and to find out its translation equivalence in Hindi. The most popular existing online English-to-Hindi translation systems were crosschecked with some English sentences having certain language specificity. All these systems failed to recognize the features of the input sentence and produce their translation equivalencies. At present, more than 200 anomalies of the existing online English-to-Hindi translation systems have been identified. It has also been noted that the functional words like adpositions and auxiliary verbs, etc,. are the most problematic lexical items in English-to-Indian languages MT [6]. Adpositions in English are termed as prepositions. There are many types of adpositions found in all languages [6, 18]. From these, the Simple English Prepositions (SEPs) are the most frequently used as well as most problematic lexical category that come under NLP applications, particularly in an English to any other languages MT context. All the SEPs have one-to-many translation equivalences in Hindi [6]. Finding the translation equivalences for SEPs with their equivalents in target languages according to the semantics of the complete sentence remains a difficult task for the machine [16]. Thus, this study concentrates on the sense identification of the SEPs and their transfer into Hindi with Machine-Learning (ML) approach.

English auxiliary verbs '*be*' and '*have*,' modal auxiliary verb '*make*,' and helping verb '*get*' have the capacity to perform different grammatical functions. Considering their grammatical functions, they have different translation equivalences in Hindi [7]. These verbs are used to express the causative sense of the main verb in the sentence. The causative sense expressed with the help of a causative verb is called Periphrastic Causative Constructions (PCC). In Hindi, the inflected forms of the main verb itself express the causative sense of a verb. The main verb undergoes morphological variations when it needs to express causation in Hindi. The unaccusative verbs possess one causative form and the accusative verbs in Hindi have two different causative forms; therefore, the selection of the causative form of the main verb form in Hindi needs special attention. For the transformation of the periphrastic causative construction from English to Hindi, a combination of a Rule-based cum knowledge-based Machine-Learning (ML) translation system is developed.

2 Knowledge-Based Machine-Learning Approach for the Transfer of Simple English Prepositions into Hindi

Translation is a linguistic operation done over the material of two natural languages. The linguistic material that is to be translated is called the Source Language (SL) and the language of the translated material is called the Target Language (TL). In the process of translation, the SL linguistic elements are transformed into the TL linguistic elements without losing the essence and aesthetics of the SL material or the SL text. The essence and the aesthetics of a text depend on the expressiveness of its lexical items and the way of presentation of the lexical items. Each word in any natural language has its own identity, and at the same time, they are related to each other. Each word in a lexicon is interconnected in many ways:-

1. Through their physical similarity/similarity of their forms–morphological similarity
2. Through their function – functional similarity
3. Through their senses – semantic similarity

Considering the functional similarity, two types of words are found in the lexicon of all natural languages: *content* words and *functional* words. The words that form the Parts of speech (POS) category like verb and noun are the content words, and the lexical items like adpositions, auxiliary verbs, and modal auxiliary verbs are the functional words. Both the content words and functional words of the lexical items are highly expressive. In a lexicon, each individual word possesses individual meaning. Adpositions are one of the functional word categories in all natural languages [7, 16]. When compared to other word categories, they are very few in number, but they play a crucial role in the formation of a meaningful sentence in a natural language. Adpositions show the semantic relation between the nouns and the verb in a sentence. In a sentence, they always come together with a noun. In some languages, it precedes the noun, and in some other languages, an adposition follows the noun. Considering their position with respect to nouns, they are called pre-/postpositions. In English, they always precede the noun, so they are called prepositions, and in Hindi, they follow the noun so they are called postpositions [9, 12]. A pre/postposition with a meaningful syntax always takes a noun as its object and forms a Prepositional Phrase (PP) as *"he relied on the weather," "the translation of the book"* [16]. In English, same and different Simple English Prepositions (SEPs) are capable of producing spatial and temporal relations between the named entities of the sentence. It must be noted that the same Simple English Preposition (SEP) produces various senses of temporal and spatial relations. The conceptual temporal sense of prepositions is determined not only by its object (P-object) but by the preceding named entity also. Therefore, this study proposes that the SEPs have no meaning, but the actual sense of the preposition in a syntactic construction is induced into the preposition by its co-occurring lexical items or the content words (verbs and the preceding and the following nouns). This postulation leads to the idea of building a knowledge base for the Word Sense Identification (WSI) of Simple English Preposition, and this knowledge base is utilized for the sense identification and transfer of the SEPs into Hindi by supervised Machine-Learning (ML) approach [19].

For the transfer of Simple English Preposition '*at*' in an English-to-Hindi MT context, we first utilize the data-based ML approach. Supervised Vector Machine (SVM) is applied on the data. But the results of the experiments are not up to a satisfactory level. So, for the transfer of Simple English Preposition '*to*' and '*with*,' we applied knowledge-based supervised machine-learning approach. In this experiment, we utilized the information of the verbs in the concerned sentences using Levin's verb classification and the traditional noun class information of the noun in the concerned Prepositional Phrase (PP). Support Vector Machine (SVM) is one of the most commonly used machine-learning algorithms. SVM is a linear classifier, which learns the most suitable hyper plane from the data [2]. The objective function for linear SVM in primal form is defined as follows:-

$$\min_{w,b} \frac{1}{2}\|w\|^2 + c \sum_{i=1}^{n} \xi_i \tag{1}$$
$$st : y_i \left(w.x_i - b\right) \geq 1 - \xi_i \quad \xi_i \geq 0$$

Here, x_i, y_i $i = 1, 2, \ldots, N$ represents the data points and corresponding class labels, w is the weight vector, e is the error, and c is the control parameter. In order to perform the experiment, the following feature representation is used to denote the data:

1. Vector representation for the sentence
2. Vector representation for the preposition under study
3. Vector representation for the verb
4. One-hot encoded representation for the verb class
5. Vector representation for the noun
6. One-hot encoded representation for the noun class

For the experiment, the control parameter is fixed as 1 and the kernel chosen is "liner" (Table 1). Table 2 shows the parameters used for the experiment using SVM.

Measures are obtained with various cross-validations. It can be observed that for the experiment related to preposition "*with*" and "*to*," a competing score is obtained for 7 fold cross-validation (Table 3).

Table 1 Data statistics [8]

	Sense	Data
with	S1	100
	S2	100
	S3	100
to	S1	100
	S2	100
	S3	100
	S4	100

Table 2 Parameters for the experiment [8]

Parameter	Parameter value
Kernel	Linear
Degree	1
C value	1
Gamma	0.001

Table 3 Performance analysis of sense identification of SEPs '*with*' and '*to*' [8]

SEP	Cross validation	Accuracy score (%)	Precision	Recall	F1 score
with	10	69.48%	0.67	0.69	0.67
	7	71.43%	0.74	0.71	0.68
	5	67.53%	0.67	0.68	0.64
	2	61.36%	0.60	0.61	0.61
to	10	77.70%	0.78	0.78	0.77
	7	76.21%	0.77	0.76	0.76
	5	73.61%	0.75	0.74	0.73
	2	66.17%	0.65	0.66	0.65

Table 4 Classification report
5 fold cross-validation of SEP
'*at*'

Labels	Precision	Recall	F1 score	Support
par	0.97	0.98	0.98	267
mein	0.95	0.93	0.94	109
ko	1.00	1.00	1.00	101
Avg/Total	0.97	0.97	0.97	477

Table 5 Classification report
5 fold cross-validation of SEP
'*by*'

Labels	Precision	Recall	F1 score	Support
baje	1.00	1.00	1.00	98
ko	1.00	1.00	1.00	266
Avg/Total	1.00	1.00	1.00	364

For the sense disambiguation of preposition, this study considered the linguistic features of the whole predicative part/Verb Phrase (VP) of the sentence. The main verb, auxiliary verb/verbs, direct object, and indirect object/objects are the main components of a predicative part/VP of the sentence. Here in this study, Direct Object is termed as predicative noun/NP1 and the Indirect Object is termed as Prepositional object (P-object)/NP2. Getting the correct sense expressed by the preposition is very essential to identify the class of the predicative noun (Perdi-N/NP1/DO) and Prepositional object (P-obj)/NP2.

During the study related to the transfer, the senses like comitative and associative sense of SEP '*with*' the PNG information and class of predicative noun/NP1 of the Verb Phrase (VP) has a vital role to determine its equivalence in Hindi.

In traditional English grammar, the nouns are grouped as common nouns, countable or uncountable nouns, and abstract nouns. If the predicative noun and prepositional object comes under the same classification, it creates confusion for the machine to select the exact translation equivalent in Hindi. To solve the issue, this study used a new noun classification method to distinguish between the predicative noun and the prepositional noun/object. For the sense identification of Simple English Preposition '*by*,' this study follows Levin's verb classification and the new noun classification method in the classification of the nouns and obtained a quality output,the results of the experiments shown in Tables 4 and 5 and the Fig. 1.

3 Transfer of English Periphrastic Causative Constructions into Hindi Morphological Causative Constructions

The syntax and the semantics of a particular type of sentence is determined by the Main Verb (MV) and the Verb Phrase (VP) in that sentence. Syntactically, verbs classified into three categories: intransitive verbs, transitive verbs, and ditransitive verbs. Semantically, verbs classified into three categories: action verbs, state verbs, and process verbs. Typology of the verb and the pattern of Verb Phrase (VP) in a sentence express certain language specific features. Causative sentences are one

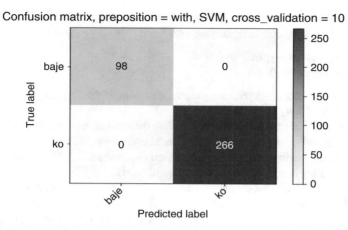

Fig. 1 Confusion matrix of temporal sense of SEP '*by*'

of these categories. The language specific features of causative sentences vary from language to language. The idea of causation is a part of the semantics of the verb itself. The causative constructions show semantic relation between the verb and the subject, Direct Object (DO), Indirect Object (IO), and Oblique Object (OBL) [4]. Causative sentences mean that one named entity (NP1) makes another named entity (NP2) do something or causes it to be in a certain state. Martin Haspelmathipzing states that a specific verb form or construction that denotes a situation, which contains a causing subevent and a resulting situation [11]. There are three types of causative constructions found in different languages. They are-lexical causatives, morphological causatives, and periphrastic causatives [3, 4, 7]. The considered language pair English and Hindi from two different linguistic and socio-cultural backgrounds, and hence, all these differences are vivid in the expression of causation also. In English, these three types of causatives are identified. The English Lexical and Morphological Causative Constructions (Eng-LCC and Eng-MCC) can be equated to Hindi with its corresponding transitive verb forms, but the English Periphrastic Causative Constructions (Eng-PCCs) are equated into Hindi Morphological Causative Constructions (Hin-MCCs).

3.1 Defining Causatives

All Natural Languages (NL) use various linguistic techniques to convey the sense of causation. The sense of causation is a basic human nature, so the linguistic expression of causation is very common. Linguistically, this expression is called causative construction/causative situation. A causative construction has a distinct syntax; apart from its syntax, a causative sentence conveys some extra-linguistic elements– a causing situation/causing event and a caused event [3, 10]

 (i) a causing event (e_1) and
(ii) a caused event (e_2)

a causative situation expressing some extra-linguistic relation between two subevents/situations [10, 11, 14]:

The net result of the two subevents reveals the causative event or situation. It can be represented diagrammatically 'E' as the causative situation or event, 'e_1' as the causing event/situation, and 'e_2' as the caused event/situation. The relation between 'e_1' and 'e_2' may be connected directly or indirectly.

In other words, we can define a causative situation as the causal verb indicating the causing of one named entity to do something, instead of doing it by itself. Semantically, causative verbs refer to a causative situation, which has two components: (a) the causing situation or the antecedent (e_1) and (b) the caused situation or the consequent (e_2). A causative construction is a construction that conveys the meaning 'an event causes another event.'

1a. John made the children dance [15]

In this example, one named entity '*John*' (NP1) did something (event e_1), and because of the event e_1, the other named entity 'the children' (NP2) as the Direct Object (DO) were compelled to perform some action event e_2 (dance). We can paraphrase 1a as:

1b. John forced the children to dance [15]

From this example, it is vivid that the real subject (NPI) does not perform the action itself, but causes someone or something else to do it instead of doing by self [22]. From 1a and 1b, it is clear that the causing event (e_1) which is not clearly specified but is indicated by the causative verb/inflected/causativized form of the main verb. Considering these facts, a causative construction possesses three important properties:

 (i) There is a linguistic element that denotes an event, which can be called the causing event or the antecedent (e_1)

Fig. 2 A causative event

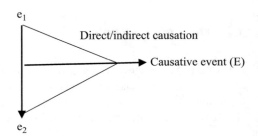

(ii) There is another linguistic element that denotes another event, which can be called the caused event or consequent (e_2) [13]

(iii) There is another linguistic element that denotes the relation between the event one (e_1) and event two (e_2)

These three combine to make a causative situation (E). Causation has a temporal dimension [15]: the causing event (e_1) must precede the caused event (e_2). A causative Construction has four main components: the *CAUSER*, the *CAUSEE*, and *the PATIENT/*AFFECTEE the EFFECT [13, 20]. In a Causative Construction (CC), these four components have their own grammatical role to determine the syntactico-semantics of the causative construction (CC) [5, 21].

1. **The CAUSER:** The causer in a causative sentence is the entity viewed as causing the entire event, i.e, the subject of the sentence [5, 13, 20, 21].
2. **The CAUSEE:** The causee is the entity carrying out the activity designated by the effected predicate [13, 20, 21].
3. **The PATIENT/AFFECTEE:** The Patient/affectee is the entity that is the endpoint of the energy (literal or metaphorical) expended in the entire causative event [12, 13].
4. **The EFFECT:** (Event E) is the resultant effect of e_1 and e_2, is the main verb (MV) in the causative construction.

3.2   Typology of English Periphrastic Causative Constructions (Eng-PCCs)

In a Periphrastic Causative Construction, the predicate of causation has certain parameters. The parameters are called *direct* and *indirect* causation. Considering these two parameters, English has two types of periphrastic causative constructions (PCCs): the direct causation and the indirect causation.

1a. She made it fall over. [20]
1b. The sun made the flowers wilt. [17]

From the examples 1a and 1b, the causer acts on the causee without any intervening entity [20]. These types of causative constructions are called direct causations (Fig. 3).

In an Eng-PCC, if the speaker/ real subject/ causer of the action does not want to reveal the real performer/ CAUSEE of the action, it is considered as

Fig. 3 Syntax of direct/impersonal causative construction

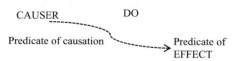

Fig. 4 Syntax of
indirect/interpersonal
causative construction

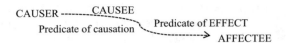

impersonal causative sentence. The speaker/real subject/causer simply mentions the
AFFECTEE/patient/DO and the predicate of effect.

1c. I *had* my car repaired.
1d. I *had* the student sit down.

In the examples given as 1c to 1d, English Periphrastic Causative Construction
have the elements such as a CAUSER, a causative verb, a CAUSEE, and a
patient/AFFECTEE, but in 1c & 1d, '*the CAUSEE*' is not expressed in that construction. This type of PCC is termed as impersonal periphrastic causative construction
(im-PCC). The pattern of the impersonal periphrastic causative construction (im-
PCC) is subject + causative-verb + object + nonfinite-complement. The syntax
of an impersonal periphrastic causative construction (im-PCC) is CAUSER +
Causative − HAVE/GET/MAKE + DO + Predicate of EFFECT.

The impersonal causative sentence possesses the same syntax of the direct
causative sentence. If the speaker/real subject/CAUSER of the action wishes to men-
tion the real performer/CAUSEE of the action and the speaker/real subject/CAUSER
also mentions the AFFECTEE/patient/DO and the predicate of EFFECT, that kind
of Eng-PCC is considered as an interpersonal causation. In a Lexical Causative
Construction (LCC), the physical involvement of the CAUSER/real subject is
vivid and that type of causative construction expresses the state of change of the
AFFECTEE/PATIENT [13, 20] (Fig. 4).

1e. I *had* a *mechanic* repair my car [3, 4, 12].
1f. We *made* the *children* clean up the playroom.
1g. I did not have *Bill* write the article [3, 12].

When compared with impersonal Periphrastic Causative Construction (im-PCC),
in the examples 1e-1g, the real performer of the action or the CAUSEE is
mentioned. This type of PCC has two agents: the first AGENT and the second
AGENT. First AGENT is the CAUSER and the second AGENT is the CAUSEE
of the PCC. This type of sentence is called interpersonal Periphrastic Causative
Constructions (in-PCC). The syntactic pattern of Eng-PCC is CAUSER+causative-
HAVE/GET/MAKE+CAUSEE+Predicate of EFFECT+AFFECTEE.

3.3   Translation equivalence for Eng-PCCs into Hin-MCCs

Before developing a particular MT system for Eng-PCCs, this study identified
the syntax and semantics of positive, negative, and interrogative constructions of
English periphrastic causative sentences and their translation equivalences in Hindi.

3.4 Transformation of English Periphrastic Causative Constructions (Eng-PCCs) into Hindi Morphological Causative Constructions (Hin-MCCs) in the context of English–to–Hindi Hybrid MT system

This system is designed only for the transformation of English Periphrastic Causative Construction (Eng-PCC) into Hindi Morphological Causative Constructions (Hin-MCCs) part of the sentence is handled by the rule-based module. The temporal prepositional phrase in the sentence is handled in two ways–the temporal prepositions '*at/by*' are identified and transferred into Hindi with ML method of translation and the P-object the temporal noun is identified and translated into the RBMT module.

(a) *Rule-Based* Cum *Machine-Learning Method of MT System for the Transfer of English Periphrastic Causative Sentences into Hindi*

The proposed system has two main components:-

1) The Rule-based system for the translation of Eng-PCCs into Hi-MCCs
2) SVM for the WSI of temporal SEPs 'at' and 'by'.

3.4.1 Overall System Architecture

(a) *Rule-Based* Cum *Machine-Learning Method of MT System for the Transfer of English Periphrastic Causative Sentences into Hindi*

The English Periphrastic Causative Constructions (Eng-PCCs) are transferred into Hindi with a Rule-based MT system. The system first identifies whether the input sentence is a causative one or not with the help of syntactic parsing; then, the system recognizes if the causative sentence is impersonal or interpersonal. After completing this step, machine selects the appropriate causative verb form in Hindi according to the tense aspect and the number and gender of the subject/object. The system translates source English words with the help of bilingual dictionaries. The system then rearranges the translated lexical items. The subjective, objective, and instrumental case markers are selected and arranged on their locus with the Paninian perspective. The Simple English Prepositions outside the periphrastic shell can be translated with the help of Machine-Learning (ML) method of translation. At the time of machine/human translation, the selection of causative verb form in Hindi depends on many factors; these factors determine the linguistic features of both the source and the target languages. All the existing English–Hindi MT systems are unable to identify the causative role of the auxiliary verbs like '*have*,' '*get*,' and '*make*' in a verb periphrasis. Rule-based MT system is a classical method. Here, this study considers the linguistic features of both the source and the target language. This system possesses a balanced translation grammar of English and Hindi, which can translate the English PCCs into corresponding Hindi equivalents. Figure 5 flow chart explains the methods of the translation process and the selection of the translation equivalence for the English periphrastic causatives into their corresponding Hindi equivalents.

(b) *The Sense Identification and Transfer of Simple English Prepositions into Hindi with ML Approach*

A knowledge-based supervised machine-learning approach is utilized to the sense identification and transfer of the Simple English Prepositions (SEPs) into Hindi (Fig. 6).

3.4.2   Experiments and Results

In order to get the specific linguistic patterns, 200 English source sentences of each type of sentences are parsed and the sentence patterns of the English causative sentences are prepared. After preparing the data sets, a chosen chunk of 100 English periphrastic causative sentences from different categories are manually translated into Hindi. By this, the translation equivalences of Hindi Morphological Causative Constructions (Hin-MCCs) for the English Periphrastic Causative Constructions (Eng-PCCs) are identified.

After this, a set of rules are prepared for transferring the English periphrastic causative sentence patterns into Hindi morphological causative patterns. These rules are utilized to implement the translation process. Rules are prepared separately for the English interpersonal and impersonal causative sentence patterns. After implementation, it has been observed that the English impersonal causative sentences with '*causative-have*', '*causative-get*' and '*causative-made*' give exact Hindi translation output compared to other causative sentences. The reason is that the other causative sentence types depend heavily on the interpretation of Simple English Prepositions, tense and gender inflections of Hindi verbs, and so on. In certain contexts, open source parser parsed certain words belonging to noun category; as in adjective category, this also affects the accuracy of the system. The newly proposed system is verified with 200 English sentences from each category. From the output of the newly proposed system, 100 sentences of each category of sentence are selected for human evaluation. The evaluator evaluated these selected sentences using different factors (adequacy and fluency) [1], which determine the quality of the MT output. The maximum score is 5 and the minimum score is 1.

After getting the scores for each sentence, we find out the average scores of each category of sentence (Fig. 7).

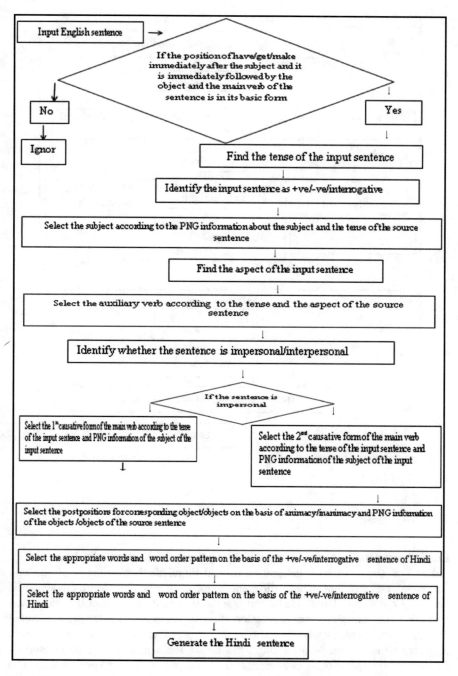

Fig. 5 Flow chart for the translation process of English periphrastic causative sentence into Hindi [7]

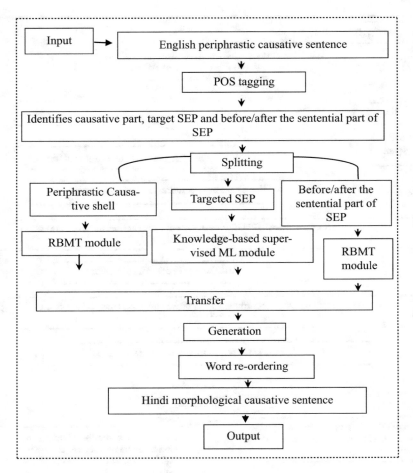

Fig. 6 HMT architecture

4   Conclusion

English and Hindi are two popularly by spoke as well as official languages of India. The socio-cultural, geographical, and linguistic backgrounds of the two languages are different. These differences are clearly visible from the semantic level of English and Hindi. Causation is a basic human nature, and it is expressed through words, and the verbal expression of causation in a natural language is called causative sentence/construction. The verbal expression of causation varies from language to language. Natural languages utilize three main linguistic techniques for the verbal expression of causation: lexical, morphological, and periphrastic. The transitive use of intransitive verb is called the lexical causation. The inflected form of the verb, which is used to express a causative situation, is called morphological

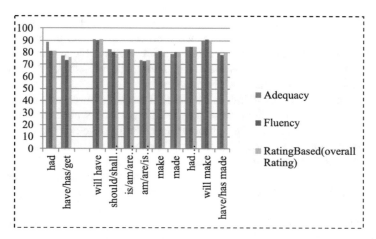

Fig. 7 Results of human evaluation of newly developed HMT system for Eng-PCCs into Hin-MCCs [7]

causative. Periphrasis is a linguistic tool, that is used to express any grammatical feature instated of inflection the verb, the Main Verb (MV) takes the advantage of some other lexical categories like auxiliary, modal auxiliary, or helping verbs etc,. These three types of causative constructions are found in English. In English, the Periphrastic Causative Construction (PCC) is two-verb construction. The first verb or the causative verb plays the formation of causative construction, and it is capable of expressing the Tense, Aspect, and Modality (TAM) of the EFFECT of the causing event e_1 and the caused event e_2. The EFFECT always participates in a causative situation in its base form or to infinitive form. English has a number of causative verbs for expressing a causative situation. Each causative verb has its own semantic specificities. Only a native speaker can specifically identify the different semantics.

This semantic specificity cannot be poured into another natural language through simple translation. English 'causative verb' like 'let', 'help', 'cause', 'force', 'get', 'have' and 'make' etc., have distinct semantics. This semantic variation cannot be translated into Hindi, because Hindi possesses only Morphological Causative Constructions (MCCs) of verb inflections. English Periphrastic Causative Constructions (Eng-PCCs) with causative-have/get/make can be equated into Hindi with morphological causative forms of the EFFECT of the source English sentences. The English Periphrastic Causative Construction (Eng-PCC) with other 'causative verbs' like 'let' and 'help' etc,. are translated into Hindi with the corresponding transitive sentence. Syntactically, there are three types of PCCs found in English; they are impersonal, the interpersonal, and the passive causatives constructions. In Hindi, two types of morphological inflections–the first causative/direct causative and the second causative/indirect causative forms–are found. English impersonal causatives are equated into Hindi with the direct causative form of the verb, and the other two causative constructions in English are equated into Hindi with the indirect causative form the verb. English has a discrete subject-verb-object word order. On the other hand, Hindi is free-word ordered language, but in the context of causative constructions, Hindi follows a strict subject-object-verb word order. In morphological causative construction in Hindi, the subjective, objective, and instrumental case markers has a well-defined role.

In the context of English-Hindi MT system, it has been observed that at present, any online-MT system cannot have handled a sentence with language-specific features of English and Hindi. This observation led to the development of rule-based MT system for the transfer of English Periphrastic Causative Constructions (Eng-PCCs) into Hindi Morphological Causative Constructions (Hin-MCCs). English Periphrastic Causative Constructions are found as part of a lengthy sentence, and the present newly developed system is only a model and can only handle the simple periphrastic causative sentence. In Hindi, the *Paninian perspective handles the kAraka relations (Case relations).* The *karaka* relations in Hindi are expressed through the Simple Hindi Postpositions (SHPs). English simple periphrastic causative constructions with *'temporal Prepositional Phrase at'* (at-PP) and *'temporal Prepositional Phrase by'* (by-PP) are handled with ML approach. The RBMT system has a balanced translation grammar of the source and the target languages. The translation module with ML approach also utilizes a knowledge base. This knowledge-base also enhances the performance of the translation model, and it is able to produce cent percent accuracy. With the help of the very strong linguistic knowledge-base of both the RBMT and ML modules–that work together, the machine produces quality translation output with cent percent accuracy.

References

1. Lavie, A. (2013). *"MT Evaluation: Human Measures and Assessment Methods"* pp 1–34. http://demo.clab.cs.cmu.edu/sp2013-11731/slides/09
2. Bojanowski, P. (2016). "Enriching word vectors with sub word information" arXiv preprint arXiv: 1607.0460.
3. Hiroshi Aoyagi (2007). "Toward a Unified Account of Morphological Causatives and Passives in Korean" In D.-W. Lee (Ed.), *Locality and Minimalism: Proceedings of the 9th Seoul International Conference on Generative Grammar* pp 9–27 Hankuk Publishing Co. August 2007.
4. John. M. Anderson (2005). "The Argument Structure of Morphological Causatives" *Poznań Studies in Contemporary Linguistics, 40*, pp 27-89.
5. Julia Horvath & Siloni, T. (2011). "Causatives across components". *Natural Language and Linguistic Theory, 29*(3), 657-704.
6. Jyothi Ratnam, D., Kumar, M. A., Premjith, B., Soman, K. P., & Rajendran, S. (2018). "Sense Disambiguation of English Simple Prepositions in the Context of English-Hindi Machine Translation System". In S. Margret Anouncia & U. Wiil (Eds.), *Knowledge Computing and Its Applications*. Singapore: Springer.
7. Jyothi Ratnam, D., Soman, K. P., Biji Mol, T. K., & Priya, M. G. (2019). "Translation Equivalence for English Periphrastic Causative Constructions into Hindi in the Context of English to Hindi Machine Translation System". In R. Kumar & U. Wiil (Eds.), *Recent Advances in Computational Intelligence. Studies in Computational Intelligence* (Vol. 823). Cham: Springer.

8. Jyothi Ratnam D., Soman, K. P., Premjith, B., & Priya, M. G. (2018). "Transfer of Simple English Preposition 'to' and 'with' into Hindi Utilizing Linguistic Features of the Predicative part of a Sentence with Machine Learning Approach in an English to Hindi MT Context". *Journal of Advanced Research in Dynamical & Control Systems, 10*, 240–264.
9. Kamata Prasad Guru. (1993). Hindi Vyakarana. *Nagarani pracharasabha* (17th ed.). Varanasi.
10. Kağan Büyükkarci, & Duygu İşpinar (2013). "A study on the Acquisition of English Causatives by Turkish Learners of English". *Cumhuriyet International Journal of Education-CIJE, 2*(4), 63–71.
11. Liesbeth Degand (1994). "Towards an Account of Causation in a Multilingual Text Generation System" In *7th International Generation Workshop*, Kennebunkport, Maine, June 21-24, 1994, pp 108-116.
12. Haspelmathipzing, M. (2008). "Causative and anti-causatives" In *Leipzig Spring School on Linguistic Diversity*, March 2008, pp 1-15. Available: https://www.eva.mpg.de/lingua/conference/08_springschool/pdf/course_materials/Haspelmath_Causatives.pdf
13. Auersperger, M. (2012). *"English Causative Constructions with the Verbs have, get, and make, and their Czech Translation Counterparts"* Available: https://dspace.cuni.cz/bitstream/handle/20.500.11956/42425/BPTX_2010_1__0_274853_0_93126.pdf?sequence=1&isAllowed=y
14. Jain, P. (2016). From "Pre-position" to "Post-position" *International Journal of Modern Computer Science (IJMCS), 4*(5). ISSN: 2320-7868 (Online) Volume 4, Issue 5.
15. Nadathur, P. (2017). *"Causative Verbs Introduction to Lexical Semantics"* Available: https://web.stanford.edu/~pnadath/handouts/ling130b-fall17/Nadathur-analytic-causatives.pdf
16. Malik, P., Gupta, A., & Baghel, A. (2013). "Key Issues in Machine Translation Evaluation of English-Indian languages" *International Journal of Engineering Research & Technology (IJERT), 2*(10). ISSN: 2278-0181, Volume 2, Issue 10.
17. Chatti, S. (2012). "The Semantic Network of Causative MAKE" *ICAME Journal, 35*, 1–14. Available: https://www.semanticscholar.org, Papers the semantic network of causative MAKE
18. Mätzig, S. (2009). *"Spared Syntax and Impaired Spell-Out: The Case of Prepositions in Broca's and Anomic Aphasia"* Submitted for the Degree of Doctor of Philosophy, Division of Psychology and Language Sciences, University College London.
19. Soman, K. P., Loganathan, R., & Ajay, V. (2009). *"Machine Learning with SVM and other Kernel Methods"* New Delhi: PHI Learning Pvt. Ltd.
20. Kemmer, S., & Verhagen, A. (1994). "The Grammar of Causatives and the Conceptual Structure of Events" *Cognitive Linguistics*, Volume 5, Issue 2, 115-156. Available: https://pdfs.semanticscholar.org/8cbb/3d8c146f1c17e826f12aa5a0bfca292a5c43.pdf.
21. Sugawara, T. (2016). "The Syntactic and Semantic Structures of the English Causative to-Infinitives" *Interdisciplinary Information Sciences, 22*(1), 57–79. ISSN 1340-9050 print/1347-6157 online http://dx.doi.org/10.4036/iis.2015.R.03.
22. Walid Mohammad Amer. (2010). *Causativity in English and Arabic: A contrastive study.* Available: http://site.iugaza.edu.ps/wamer/files/2010/02/Causativity-in-English-and-Arabic-latest.pdf

Index

A

A4A, *see* Access for All (A4A)
Access for All (A4A)
  AChecker, 23, 25
  Tenon tool, 25, 27
  WAVE, 25, 26
Accessibility, 23, 24
Accessibility Guidelines Working Group
  (AGWG), 19
Accessible tourism
  CRPD, 16
  elements, 16
  and hospitality organization, 14
  ICF, 16
  PwDs, 16
  social integration and economic condition,
    16
AChecker, 14, 17, 21, 23, 25, 28
AC induction motors, 77
Acrylonitrile butadiene styrene (ABS)-based
  filament, 75
Acute lymphoblastic leukaemia (ALL), 108
Adequacy, 242
Adjectives
  antonymy, 214
  descriptive, 214
  gradation, 215
  markedness, 216
  similarity relation, 214, 215
Adpositions, 250
Adverse drug events (ADEs), 83
Agglutinative language, 227
Agriculture, 59
Alexa, 15, 23, 28

Alzheimer's disease, 93
Ambiguity
  category, 139–140
  definition, 134
  lexical, 135–139
  structural, 134, 135
Annotation decisions, 152–154
Antonymy, 206, 214, 219
Application processing layer, 94
Applications of CDSS
  ANN (*see* Artificial neural network (ANN))
  Bayesian network, 91–94
  Belief rule-based architecture system,
    94–96
  BRB architecture system, 94–96
  evidence-based, 96–97
  fuzzy logic, 86–91
  genetic algorithms, 106–109
  hybrid systems, 103–105
Arduino Pro Mini Microcontroller boards, 66
Arduino UNO microcontroller board, 61
ARM 7 Processor, 61
Artificial Intelligence (AI), 59, 247
  applications
    CDSS, 40–41
    clinical decision support systems, 40
    drug development, 41–42
    IBM Watson Health care, 40, 41
    in medical industry, 39
    radiology, 41
    single computer system, 40
    telemedicine, 41
  biometric logins, 52
  cloud architecture, 45–46

© Springer Nature Switzerland AG 2021
R. Kumar, S. Paiva (eds.), *Applications in Ubiquitous Computing*, EAI/Springer
Innovations in Communication and Computing,
https://doi.org/10.1007/978-3-030-35280-6

Printed in the United States
by Baker & Taylor Publisher Services